RED ROCKS COMMUNITY COLLEGE

U18960 043 607 9

D0744220

TA 1800 .S24 1988

Safford, Edward L.

Fiberoptics and laser
 handbook

DATE DUE

MAR 28 '91			
AUG 8 '91			
FEB 27 '92			
NOV 2 5 '93			
DEC 1 5 '93			
APR 19 '94			

DEMCO 25-380

FIBEROPTICS AND LASER
HANDBOOK
2ND EDITION

Edward L. Safford, Jr. and John A. McCann

RED ROCKS
COMMUNITY COLLEGE

TAB **TAB BOOKS**
Blue Ridge Summit, PA

My comprehensive appreciation to those who assisted me in the writing of this book, which is dedicated to my daughter Kathleen.

—John A. McCann

To all students both young and old who can imagine the growth of technology and who bring forth the new ideas and inventions that constantly make our world a better and nicer place to live.

—Edward L. Safford, Jr.

SECOND EDITION
THIRD PRINTING

Printed in the United States of America

Reproduction or publication of the content in any manner, without express permission of the publisher, is prohibited. No liability is assumed with respect to the use of the information herein.

Copyright © 1988 by TAB BOOKS

First Edition copyright © 1984 by TAB BOOKS

Library of Congress Cataloging in Publication Data

Safford, Edward L.
Fiberoptics and laser handbook.

Includes index.
1. Fiber optics. 2. Lasers. I. McCann, John A.
II. Title.
TA1800.S24 1988   621.36′92   87-35676
ISBN 0-8306-2081-8
ISBN 0-8306-2981-5 (pbk.)

TAB BOOKS offers software for sale. For information and a catalog, please contact TAB Software Department, Blue Ridge Summit, PA 17294-0850.

Questions regarding the content of this book should be addressed to:

Reader Inquiry Branch
TAB BOOKS
Blue Ridge Summit, PA 17294-0214

Cover photograph courtesy of Coherent, Inc., Palo Alto, CA.

Contents

RED ROCKS
COMMUNITY COLLEGE

Introduction

LASERS ARE RAPIDLY BECOMING STANDARD EQUIPMENT FOR MOST AUDIOPHILES as the compact disk player and laser disk become more readily available and affordable. The future holds much promise for the application of the laser in many services of mankind. As this instrument develops from the laboratory stage to the homeowner, most of us will need to be educated to the practical side of the laser. One of the most promising uses for lasers is communications over fiberoptic cable as seen with the advent of new long distance telephone companies relying totally on this technology.

The miracle of optical fibers is permitting more information, intelligence, and data to be transferred from one point to another more quickly and precisely than ever thought possible. As their algorithms and equations are developed and proved, fiberoptics shall become as common as wire, easy to construct to precise tolerances, accurate and indefatigable in operation, and a type of circuit which enhances our lives and ways of living.

It is good to know about and understand fiberoptics and lasers even though you may not be able to experiment with them for lack of fiberoptic material and all the necessary experimental equipment. Time will solve this inconvenience. Strands of optical fibers used in various capacities (from kits for the younger experimenters, to various devices using optical fibers for the more advanced) will become very common in the future. More and more uses for these solid-state devices are becoming available to us.

Chapter 1

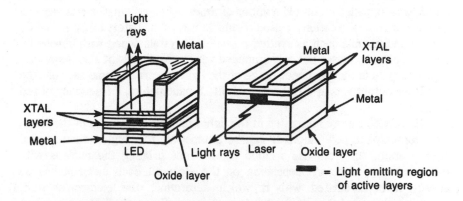

Modern Physics and Laser Technology

ALASER PRODUCES OPTICAL RADIATION USING A POPULATION INVERSION TO provide Light Amplification by Stimulated Emission of Radiation. Laser radiation may be highly coherent either temporally, spatially, or both.

WHAT IS THE BASIC PRINCIPLE OF LASING?

Light amplification was once accomplished by using a series of converging and diverging lenses on the most powerful source available, the sun. Aside from the obvious disadvantage of being tied to the sun as a source, this system was very cumbersome. The lenses required accurate placement, and the limitations on collimating the rays produced only approximated parallel rays. While searching for a better source of collimated light rays, scientists found the laser.

The laser pumps electrons to higher energy states and lets them fall back to a lower energy state while emitting rays of light. This process sounds much easier than the definition and certainly makes more sense to most people. The rays of light are coherent, monochromatic, and have linear wave fronts. Radioradar electonics enthusiasts and experts compare the laser to a single frequency continuous wave note rather than a jumble of sound produced by a spark-gap transmitter.

Radar technology gained dramatic ground when Charles H. Townes developed a device called the maser, an acronym for Microwave Amplification by Stimulated Emission of Radiation. The maser allowed the radar's range to be increased without a drastic increase in output power. Instead it minimized the internal circuit noise of the front end of a radar system. This concept had the scientific-engineering community standing on their ears. The process consisted of taking a very tiny signal and mixing it in a peculiar manner in the maser so that it was amplified an untold number of times. The early maser was either a gaseous ammonia creation or used a rubidium crystal. Later, other elements were found useable such as synthetic sapphire crystals doped with chromium (ruby). The ruby maser not only amplified very weak signals, it also generated signal outputs in the frequency range of the light spectrum. Some say that the first laser output was at a frequency of 694.3 nanometers, a frequency of red light.

Since we're more interested in the light emitting capabilities of the laser, let's contemplate how electrons can be pumped up to a higher than normal energy state. An atom has a cloud of electrons orbiting the nucleus with the number of electrons depending on the element. Each electron has an energy level associated with it which determines the electron's orbit. These energy levels are of fixed values (quanta). When an electron receives additional energy and absorbs it, the electron is forced to change orbits, going to one which requires this higher energy, leaving a hole in the electron shell. This unnatural situation leaves the atom in an excited state, extremely unstable and willing to release the unwanted energy easily. As the electrons return to their natural (ground) state, they give off this excess energy in the light beams of coherent light.

Picture an ellipsoidal chamber (which has two focal axes) with a ruby-rod material positioned along one focal axis and a helix shaped flash-tube along the other axis. The inside of the chamber is highly polished to provide all possible reflected light energy. When this flash-tube is fired using a high voltage power supply triggered by a relay closure (the flash lasts only a fraction of a second), the light energy is focused on the ruby-rod material by the geometry of the ellipsoidal chamber. Reflective mirrors are at one end and a partially silvered mirror protrudes at the other end, allowing some light to be transmitted. When properly done, a fractional-second pulse of light is emitted from the ruby-rod of such intensity that it can instantaneously burn through a steel plate! That ruby-rod was pumped-up, and there was no doubt that the output was laser light — coherent, monochromatic, and full of energy!

MORE ON PUMPING-UP

Let's delve into some familiar concepts related to the absorption of energy. A thermal switch can be composed of a bimetallic strip (two metals clad together with differing rates of thermal expansion). When this switch gains or loses sufficient energy, the switch opens or closes an electrical circuit. These thermal switches are found on ovens, water heaters, and wall thermostats. The heat energy applied seems small, but the reaction of the switch to heat is quick and powerful after enough heat has been absorbed. Usually this reaction is a physical movement.

In the laser, an absorption of energy is also accomplished, but this does not result in anything you can see or hear or feel as far as the chemical element itself is concerned. You see the result of this absorption—relaxation of the atoms in the release of the powerful laser light beams—but you cannot directly observe what happens inside the atoms as this phenomena occurs. You can see a thermal switch element begin to change shape, or expand, or whatever, if you watch it close enough and its atoms are being excited by the heat radiations to cause this physical change.

There is a model of the atom called Bohr's model. It depicts the core of a spherical unit as having a lot of balls called neutrons and protons tightly held together (by an unknown force). Orbiting around them at given distances (called shell-distances) are other smaller circular balls called electrons. Three facts are important: The neutrons have no electrical charge (neutral = neutron); the protons have a positive charge; and the electrons have a negative charge.

Even with a very elementary knowledge of electricity you know that positive charges are supposed to attract negative charges. According to that old maxim, "Nature abhors a vacuum" (or, to state it more generally, "Nature hates an unbalanced condition.") So the nature of things physical is that the positive and negative charges should not be separated. They need to get together to form a neutral pair of no charge. An electrical current flows as the electrons try to reach the protons and equalize them. Why don't the electrons in an atom fall down toward the center of the atom and neutralize the protons there? Because they are orbiting—rotating around the core with some velocity—faster as they are closer to the core, and slower as they are farther away from the core.

These electrons are like satellites around Earth. They are attracted to the earth by gravity, but they don't fall. Why? Because they are moving at such a speed that the distance they fall is the same distance that the earth curves away from the horizontal. (See Fig. 1-1 for a view of this elementary concept.) There can be an orbital path which may be circular about the earth, if the velocity of the satellite is correctly adjusted for that kind of path. An ellipsoidal path is shown, on which the satellite is fastest at perigee and slowest at apogee.

Back to the atom. The electrons in the Bohr atom are like a satellite, moving around the nucleus at a proper rate of speed so that their centrifugal force equals the attractive force of the core or nucleus, and so they continue to orbit and do not fall into the center of the atom. The electrons in the outer shell are held the most weakly in the atom by the core forces, and so under proper conditions might be caused to move out of one atom to another and then to another, etc. This is known as the "flow of electrical current." When the outer electrons "bond" with other atoms in a permanent manner then the atoms are said to have bonded together to form some kind of molecule or compound.

In the model of the atom as proposed by Niels Bohr in 1913, it was necessary to explain some discrepancies which existed in the comparison of the electron movement with that of the satellites or solar system planets as we know them. One such discrepancy was that the orbit's radius did not seem to be continuous but seemed to exist in finite steps away from the nucleus. A

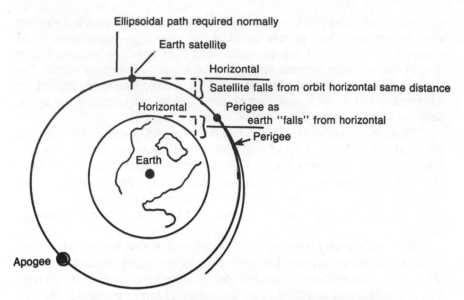

Fig. 1-1. An earth satellite in orbit is like Bohr's model of the atom.

satellite can be put into an orbit at any distance from the earth, given the proper velocity and direction of movement, but not so with the electrons. They seemed to have absolutely certain rings of orbit into which they could jump, going out further and further from the core center; they could not orbit between these acceptable rings or shells. Thus we have the concept of quantum jumps. When electrons are excited (meaning they have been given some energy in some manner and absorbed that energy) they jump into a higher orbit.

Some problems still had to be resolved. For one thing, according to classical electromagnetic theory, an orbiting electron would have to constantly lose energy by radiating light (high frequency energy related to its orbital period) and so should finally lose enough energy that it would fall into the core center or nucleus of the atom. But this doesn't happen! This led to the idea of the ground state, the distance from the nucleus at which no more energy can be given off as light rays by the orbiting electron. Bohr said that an electron would only lose energy when it jumped from a higher level of excitation to a lower level of excitation. Losing energy means that this electron emits rays of light energy; that idea is important to our concept of the emission of laser light. So Niels Bohr, genius that he was, explained the stability of the hydrogen atom and its distinctive pattern of sharply defined wavelengths of light which this atom radiates.

ELECTRON REVIEW AND ENERGY LEVELS

There is a very simple equation:

$$E = \text{Allowed Energy} = -E_0/n^2 = E_0/n^2$$

4

E_0 is a constant number and n is a positive number called the principle quantum number. This designates the energy levels of the electron. E_0 has a value in electron volts of 13.6, which is the value an electron gains when accelerated through a voltage potential of 13.6 volts. When n is equal to 1, the electron is said to be in the ground state as defined according to Niels Bohr. Thus in the ground state the hydrogen electron (a hydrogen atom has one electron and one proton and is the basis for most of these kinds of studies) has an electrical energy level of -13.6 electron volts. The minus sign shows that the force is opposite the attractive force produced by the nucleus which is positive. Higher energy levels are stated by a larger value for the letter n and in the limit, as n approaches infinity, the energy level approaches zero, as you might suspect, because then it is so far away from that core that little or no attraction exists between it and that core.

So to pump-up the energy levels of atoms in a given substance, the electrons of the atoms must somehow jump to higher orbits or into other shells by adding energy to them. The light energy added to the ruby-rod laser in the ellipsoidal chamber is one method, and there are others.

Once this extra energy has been added to those atoms in the given substance. The electrons, which are not in a normal state of orbit for that substance, won't like to stay in this excited state, and tend to jump back to their normal orbit distances. In doing this, they give off energy (light radiations). Also, since the frequency of the light depends on the orbiting frequency or rotational period of the electron around the nucleus (and in a given substance all electrons are assumed to have the same orbiting frequency) the light rays have the same frequency (are coherent). Assume that the rise to an excited level, and the return to the normal level or ground state level where no emissions take place, is accomplished simultaneously by all atoms in the substance. Then, assume that this process produces a beam of light (coherent) with much energy contained therein: a laser beam of light.

Albert Einstein theorized that light could interact with electrons so that the light itself might be considered bundles of energy or quanta. This helped explain the photoelectric effect. In this concept, light rays are considered to be particles that do not have any mass (if you can imagine that) called photons; these have a movement which is the speed of light itself. Each of these particles has an energy content that can be expressed as:

$$E_0 = \text{Photon energy level} = hf + K_{max} \text{ Joules}$$

h is Planck's constant $= 6.625 \times 10^{-27}$ erg-sec; f is the frequency of vibration or oscillation (a quantum number that is a digit and not a fraction of a digit). K_{max} represents the maximum kinetic energy that the photoelectron can have outside the metal surface.

$$\text{Joule dimensions} = \frac{\text{kg m}}{\text{sec}^2}(m) = \text{Newton-meter}$$

Kg = kilogram, m = meter.

Physics texts state that the electron has two kinds of energy. The first is energy due to movement, called kinetic energy. The second is energy due to its distance from the core or nucleus. This, like anything raised up above an attractive surface, is potential energy. The longer the fall, the larger the potential energy as long as the attractive force is present. Since electrons may not be just particles but may be "clouds of energy," and they cannot be well defined in position or orbit. It is currently popular to say that an electron cloud has a probability of being in a given space at a given movement of time. This really means that we don't know very much about the inside working of an atom even now as of this date. We know a lot about effects but since we can never observe an atom's interior, at least with instrumentation currently imagined or fabricated, we really don't know what is there or exactly how it all functions. The measuring process itself would destroy the accuracy of measurement. Some very able scientists at MIT have succeeded in separating a single atom for effects study (Klepper, Littman, and Zimmerman for their work on highly excited atoms, Scientific American, May 1981). There may be hope that in the future still more exciting developments to astound classical physicists will occur.

A SMALL REVIEW OF PLANETARY-SATELLITE CONCEPTS

A review of satellites and our own solar system and the planets revolving around the sun will help with further understanding of the inside of atoms.

Two masses attract each other with what is called a gravitational force. This force can be accurately measured and is, in the centimeter-gram-second system of units for the attraction between two masses, each of one gram weight, that are one centimeter apart:

$$G = 6.673 \times 10^{-8} \frac{\text{dyne cm}^2}{\text{gm}^2}$$

(or about one 15 millionth of a dyne).

The equation for the attraction of a satellite to the earth is:

$$\text{Attractive Force (gravitational)} = G \frac{Mm}{R^2}$$

M = mass of earth, and R = distance between the center of mass of the earth and the center of mass of the satellite.

The force varies inversely with the distance between the two objects, the earth and the satellite mass.

There is another equation relating to the velocity of the object in orbit. It is:

$$\text{Velocity required} = V = \sqrt{R_0 G}$$

(V) is meters/seconds, R_0 is earth radius in meters and G is the gravitational constant in the appropriate units of meters-kilograms and seconds. (Of course, the British units can be used so long as the units are consistent.) The velocity stated here is the speed required to maintain an object in a circular orbit at the earth's surface.

The equation can be stated in another way:

$$\frac{mv^2}{2} = G\,\frac{Mm}{2R_1}$$

Where R_1 is the distance from the earth's center to the satellite orbiting above the earth. This equation is quite interesting, as it shows that the kinetic energy mv^2 needed to maintain the satellite in orbit decreases as the satellite gets farther from the earth, just as the electrons in an atom. The further out they are from the core, the weaker they are held or bound to the core. However, the potential energy (PE) of the satellite is related to its distance from the earth by the expression:

$$PE = -mG\,\frac{R_0}{R_1}$$

R_0 = radius of earth, R_1 = satellite's orbit radius to center of earth, m = satellite mass, and G = gravitational constant.

At an infinite distance from the earth the PE is zero, since the satellite is so far away it is no longer attracted to the earth. So it is with the electron mass-cloud. When it is far enough away, it is not attracted to a given nucleus of a given atom.

WHAT HAPPENS IN THE ATOM?

Let's assume that the electon is a cloud of energy and that a photon comes along and collides with it. What seems to happen is that the electron absorbs the photon and increases its own energy level. It gobbles up the photon like Pac-Man gobbles up energy pellets. If its own energy level is increased enough, it moves farther away from its nucleus to another shell or ring or orbit and is considered to be in an excited state. What that implies is that its energy level is increased, but only for a short period of time. It cannot retain that new energy level. Therefore, it must give off some energy (of a different kind perhaps) so it can return to its stable and normal energy level condition. What it gives up is a light ray, which also may be considered a photon under some circumstances. Tests show that the electron-cloud may give up just one photon as it returns to its original orbit, or it may give up several photons through a series of intermediate permissible orbits. It cannot orbit just anywhere. It can only orbit in integer unit levels.

It is convenient to think of the electrons in an atom as being on the steps of a staircase as illustrated in Fig. 1-2. The steps are not equally spaced; more

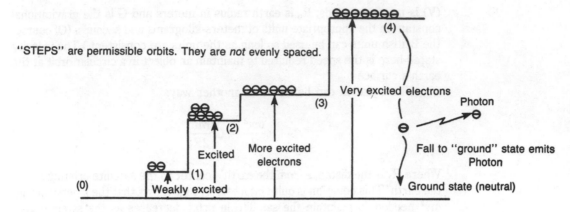

"STEPS" are permissible orbits. They are *not* evenly spaced.

Excited

(2)

Weakly excited

(1)

Excited

(0)

More excited electrons

(3)

Very excited electrons

(4)

Photon

Fall to "ground" state emits Photon

Ground state (neutral)

Fig. 1-2. "Excited" electrons.

than the usual number of steps (3) are shown for clarity. Consider that an electron may fall from the highest step to the one below it (from step 3 to step 2) and emit a burst of energy and then fall to the ground state. This means that, when pumped-up, the electrons are given the energy sufficient to move them from an orbit that is closed to the nucleus to an orbit they usually would not have. They give up that added energy and drop back to the neutral (ground) state orbit.

An "inverted" population refers to the number of electrons present at some particular time in a given orbit, more than would normally exist in that shell at that distance from that nucleus. Inversion usually means to turn something up side down. Here it means to have an overabundance, or more than usual.

Another point worth clarifying is the name maser (Microwave Amplification by Stimulated Emission of Radiation). A microwave frequency is generated or amplified by this process. The word laser changes the frequency from microwave to that of light; Thus we have Light Amplification by Stimulated Emission of Radiation. The only difference is the frequency of output being generated.

Atoms are very complex. An atom is excited by gaining energy either by collision with another atom or by absorbing photons of light (and many astronomers think this absorption is of infrared light frequencies). The energy level of an atom so excited must increase by a quantum jump to the next (or many steps) higher in the energy staircase shown in Fig. 1-2. Since the stairsteps are steeply fronted, the electrons or excited atoms cannot exist anywhere except on the steps or at discrete levels above the ground state. The change in energy level cannot be gradual, but occurs in quantum jumps.

When an atom or molecule drops back to a de-excited state or a less-excited state, its energy difference must be dissipated either as a photon or in a collision with another atom or molecule which then absorbs that energy.

Nature doesn't like to have unbalanced situations, inverted populations, or over-excited atoms!

Suppose some photon energy interacts with a substance causing the energy level of that substance to change. The substance becomes unbalanced and excited. When the pumping of energy is removed, the atoms of the substance seek their own balanced levels of energy. In seeking this return to the balanced level, the atoms appear to release more energy than the amount causing their excitation. If the input is light, the output is light of a stronger intensity. Likewise, if the input is some radiation other than light, the output is also a stronger radiation.

Since the energy into the system must create the excited state, you cannot get more energy out of the system than you put in. The resulting energy is focused or concentrated into a smaller spread of some dimension and hence it appears to be amplified.

There may be another way to rationalize this idea. Let's consider the word amplification as associated with the rise to an excited level and drop to a ground-state level of the electrons in a maser or laser-type device. Quantum mechanics says that atoms can only be raised to various discrete steps by energy absorption, and then they emit just exactly the right amount of energy which corresponds to the difference in the energy level to which they drop.

Many atoms go to the excited stage and then drop back simultaneously to a lower energy level that produces amplification. An interesting concept associated with this idea is that the atoms might be pumped-down to the ground states. That would mean that they have an inclination to remain in the excited state and during this pumping-down action many atoms may move simultaneously, giving an increased output called amplification.

It is possible to excite atoms and then get an increased output from them when they return to the ground state. The next section discusses how to excite them. There are large gas-filled lasers. Some lasers have solid elements in them such as synthetic rubies and integrated circuit lasers. This latter group is interesting because of the smaller size and possible use in optical-fiber application.

EXCITING THE SOLID-STATE LASER

Lasing action in elements can be caused by the absorption of photons, and that means a light intensity causing the excitation of the atoms. Also, there is a method of simply passing an electric current through solid-state, transistor-like elements which causes the atoms to become excited. In the latter, if current is provided in a proper manner and a proper geometry is provided to the transistor-like elements, windows can occur through which lasing beams are emitted.

Energy can be imparted to an atom through the process of collision. If electrons bump into one another, energy is imparted and lost in the process. This is the basis for gas-lasers using the impacts of two gasses. Certain conditions must be met before an energy transference is possible: the electrons must exist in certain energy levels around the nucleus so that a collision causes some to go to a higher state level and some losing energy to go to a lower energy state. Neither level can be the ground state. The use of radio-frequency energy can be used to excite these gaseous atoms through an rf discharge tube. This produces ionization and frees electrons to move around and cause collisions. Also, using electric currents passing through a gas can cause free electrons to jump around and collide. Once a lasing action has started, it is possible to use the laser beam itself, reflecting it back and forth in the substance or gas, to produce a build-up of the laser beam. This positive feedback arrangement is mandatory with some types of laser structures.

THE GALLIUM-ARSENIDE LASER

There are many types of commercial lasers on the market and much information can be found about them, but now let's discuss gallium-arsenide laser. It is a solid-state type and can be obtained for home experiments. It is a type that seems adaptable to optic-fiber applications.

The gallium-arsenide laser is a good example of a solid-state diode which can emit light in the infrared region when an electric current is passed through its junction. It is also called an injection laser, because when the electric current passes through the junction between the P and N materials light is emitted. This light (in the infrared region) is not visible to the human eye. The infrared spectrum consists of three regions: The NIR region or near infrared region, the IIR or intermediate infrared region, and the FIR or far infrared region. Near is near visible light frequencies and far is frequencies at the opposite end of the spectrum of infrared emissions. Injection lasers are adaptable for communications applications because they are small, light, and compact, and require small amounts of electrical energy. Their outputs are sharply focused, and they can be tuned to certain windows in the atmosphere so that absorption is relatively small. They can be switched on and off at the fast rates required for digital communications.

The windows in the atmosphere that pass infrared emissions are 0.95 to about 1.00 microns, 1.2 to 1.3 microns, 1.5 to 1.8 microns, 2.1 to 2.4 microns, and 3.0 to 5.0 microns. These are not exact, but are close enough to give us some idea of what the window frequencies are. There are other frequencies for other types of laser light which are windows for sea-water and so on. Weather can have an important part in determining which window is a window at what time.

Some experiments were done with a heat-seeking head for a small guided missile. I could hold a lighted cigarette some 30 feet away from the nose cone and have the control surfaces deflect as the cone tracked the cigarette. These types of missiles are sometimes used against jet aircraft and so their detectors are necessarily required to have windows which see the jet exhausts. Of course, countermeasures were developed in the form of layers which also were so designed to radiate the window frequencies strongly. When these were ejected from the target aircraft, they could present complete confusion to a heat seeking missile.

Some applications require domes over the missile sensing units. These can be of some material transparent to the window frequencies and opaque to other frequencies. Glass and quartz might be used for some frequencies and synthetic sapphires, silver chloride, and sodium chloride might be used for other frequencies.

THE STRIP DIODE LASER

A laser invention of great importance was the development of the strip edge emitter of light. In this configuration (shown in Fig. 1-3) the metal contact to the element of the diode is so arranged that it presents a strip or metallic ridge which runs through the diode at the junction of the PN materials. A section is free so that light can be emitted from the junction area. Due to the metallic connection, a very active region of electron-hole dynamics will occur in this type of lasing diode. Connection of this type of light source to an optical fiber is made using a pigtail part of the fiber.

In this strip diode-laser an initially excited photon goes down the strip of metal, and by collision and interaction with other atomic elements excites other photons. Since the ends of the strip are optically flat then reflections take place and so it is considered that the strip acts like a resonant cavity. If the oscillations are properly stimulated according to energy level and frequency, then lasing action takes place and coherent light is emitted. The two strips used in the PN junction are carefully machined so that they are perfectly flat, perfectly parallel, and reflective, and thus form mirrors so that the photons are reflected between them after initially being excited by some free electron (provided by the forward bias arrangement of external current to the diode laser). You must be very careful with this type of laser to prevent temperature increases that might destroy the unit when it is being operated. Normally there is a feedback circuit used which monitors the output and controls the exciting voltage to prevent undesired temperature increases.

The light emerging from the active area of the injection laser diode is

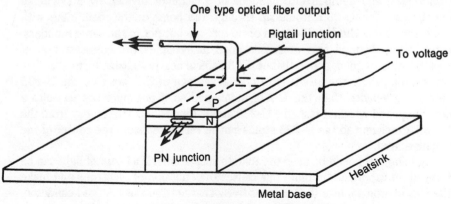

Fig. 1-3. A strip-laser light source.

coherent and is in a narrow beam whose size is governed by the active area of the diode. If you study the spectral output of such a beam you find that there is coherent light. This means that when the electrons are pumped to a higher energy level and fall to the lower levels emitting photons, they may all do this in exact time phase. Most rise and fall simultaneously, which is desired, and these produce coherent light required as an output from a laser.

It is possible to visit a radio parts store and get LEDs and request some lasing diodes. Circuits often come with these that show you how to connect them. Be careful! Never look at the output of any device that might produce a laser light. It has the power to blind you!

One problem with a lasing diode of this type has been the small amount of power available from it. Now comes the news that Xerox's Palo Alto research laboratories have developed a ten strip or stripe semiconductor laser which can emit up to 410 milliwatts of optical power. This is about seven times more than possible from the single strip (stripe) laser mentioned earlier.

The new laser is a gallium-arsenide-aluminum diode and it derives its fantastic capability from a new design with ten etched stripes (strips) instead of just one. These are all optically connected (by mirror and reflection and parallelism) so that the resultant output beam is coherent. The beam output occurs at about 825 nanometers in the infrared band. It is expected that this type of laser, if improved, and if frequency of the light output can be controlled properly, may replace the helium-neon lasers now being used in document scanners, optical data stores and systems, and on printers and so on. Most helium-neon lasers have an output of only about 50 milliwatts at this writing.

There have been laboratory experiments that are so refined that single atoms can be separated from molecules in order to study them. This concept has become an important tool in the study of atomic energy, especially the separation of U-235 from U-238. Individual atoms of these two isotopes are said to differ from each other by less than a septillionth of an ounce, or the weight of three neutrons.

Currently, in other to separate the two atoms an involved process has been required. Refined uranium is mixed with fluorine, converted to a gas, compressed and pumped into one side of a chamber divided by a sieve-like membrane. The U-235 atoms zip through the holes in the membrane with some ease while the U-238 atoms come through but not in the same numbers. After going through this process a good many times the concentration of U-235 is pretty good, above that of U-238 and in nearly a usable form. Another way of separating the atoms is to take advantage of the fact that the U-238 atoms are heavier than the U-235 atoms and so if one puts the gas into a centrifuge and whirls it fast, the U-238 atoms tend to move out away from the U-235 atoms and so the U-235 atoms can be recovered from the center of the machine area.

Lasing action can be used to separate such atoms. If a beam of light can be focused on vaporized uranium this causes the emission of electrons and make the U-235 atoms ionize (become positively charged). In this charged condition, these atoms can then be influenced by an electric field (like charges repel,

12

unlike charges attract) and so the U-235 atoms can be attracted and collected in a very pure state.

There is a catch to everything. The catch here is to have the correct laser light frequency. You can't stimulate uranium atoms with just any frequency of laser light. The scientists won't indicate the frequency of the lasing light, but they do say it is a kind of reddish color. In the machine that does the separation, a green light is used to stimulate atoms to produce this reddish light, which causes the ionization of the uranium atoms.

EXCITATION OF ATOMS

Atoms can be excited to various states by various means (light, rf energy, electrical currents, heat changes, and so on). When excited, such atoms have electrons accumulate in the outer shells so that the atom is unbalanced. They tend to return to the normal state and cause an energy emission (typically light) from the atoms. When all emissions take place in a proper timing frame you have coherent light. Atoms become more excited as more energy is absorbed and electrons move farther away from the nucleus of the atom. In the most highly excited condition, electrons may actually break away from the bonding of the atoms and drift in some manner from atom to atom. The atom they leave is then ionized, or has a net positive charge. Ionized atoms are often used to excite other atoms, being injected into, say, a chamber filled with a certain gas or gassified material by means of an injector rod. If you have had experience with radio-frequency wave generation and transmitters or television high-voltage power supplies, you probably have had the experience of seeing a blue, ionized emission, which can occur from any sharp pointed object connected to the generating source. You may have smelled an odor, kind of a burnt smell (ozone), which is released simultaneously with the arcing. That blue emission was the emission of ions, positively charged atoms, which had lost their balancing electrons.

Thus we come to the idea of highly excited atoms, or Rydberg atoms as they are called. These atoms are very large in size (for an atom), have a very long life, and are very much influenced by a magnetic field, even if it is a weak field. When highly excited, they can have an electron orbit outside the fields of other electrons, but are still held to the atom by the attraction of the core and possibly the remaining electrons. These atoms also can be explained by the Bohr Quantum Theory: i.e., atoms exist only in certain energy-level states. The electron cannot, ever, spiral into the nucleus, it can only lose energy by jumping from a higher energy level orbit to a different, lower energy-level orbit. It gives off excess energy as an electromagnetic radiation in the energy-losing process. The allowed energy level states can be described mathematically by:

Energy = E = −13.6 electron volts/principal quantum number

One electron volt is the energy gained when an electron is accelerated through one volt of potential. When the principal quantum number is 1, the level is the ground state, or has a potential of −13.6 volts. The minus means this force is opposite the nucleus. The quantum number n must be an integer. It has been found that the radius of a Bohr orbit of an electron is proportional to r^2 and for the Rydberg orbit n may be between 10 to 100, according to MIT scientists. This may be of importance in that, as Bohr surmised, the energy emissions from an electron which jump from a value of n near 100 to the next lower level from 90 to 89 are much smaller than those emissions which take place when the electron jumps from a level of n = 2 to n = 1 or the ground state. There is a smooth change in emission levels when n is large, and this contrasts with the energy emissions from atoms where the jumps are made with small values of n. This could be important in lasers which operate continuously versus those that are pulsed. In the Rydberg state, the atom is very close to being in the ionized state as that far-out orbiting electron could be torn away from the Coulombic attraction of the core at any time.

The alkali metals are the work-horses of atomic physics. They are lithium, sodium, potassium, rubidium, and cesium. They are used because they are easily converted into a gaseous state and because their spectral absorption lines are at wavelengths easily generated by laser light. The alkali atoms may be excited with pulsed, tunable lasers that can generate highly intense flashes of light for just a split second. The light, of course, is monochromatic (one frequency). Using three lasers in one experiment conducted by MIT scientists, two pulses of light got an atom to an intermidiately excited state, and another pulse from a third laser boosted the excitation such that the Rydberg state of high-excitation was accomplished.

WHEN CAN AN ATOM ABSORB LIGHT ENERGY?

An atom can absorb light when the frequency of the light, when multiplied by Planck's constant h (which has a value of 6.625×10^{-34} joule-seconds) is equal to the energy difference between the initial state of an electron and the excited state of that electron. This relationship is accredited to Bohr. It is interesting to note that when light from a lamp is passed through a gas and then spread out with a prism so the individual color frequencies can be seen, there are dark lines at those frequencies where the light wavelengths have been absorbed. This occurs if the energy at those frequencies satisfies the Bohr relationship.

In the maser, the energy level of the atoms or molecules must be raised to a particular quasistable state and then stimulate them to fall back to a lower level. The energy they give off in doing this is an amplification of the input energy. The word pumping comes from this kind of action: the raising and lowering of the energy level of atoms so they will give off energy. Pumping adds energy — a pump produces output pressure, and then draws in energy like a pump sucks in some kind of intake. It takes on the dimensions of a cyclic kind of process and electromagnetic radiations in an oscillatory state provide the right kind of energy that is required to do the pumping job.

A molecule at energy level 2 gives up that energy and, because of amplification, it takes only a very weak down stimulus to cause it to return to the energy 1 state. It gives off a quanta of energy in doing so. The quanta it gives off then triggers other atoms in the same condition to emit energy and so ultimately the energy output is millions of times larger than that which was required to get that first molecule (or atom) to return to state 1 from state 2. An atom is inverted when its energy state is like that just described. It wants to be at energy stage 1, but in some manner (radiation or collision or whatever) it was sent to level 2 or higher and it finally wound up at level 2 needing just a tiny bit of down energy to get back to level 1. Then it is no longer in an inverted state.

It would not be proper to discuss the energy states of the Bohr atom without considering an energy level diagram of the basic hydrogen atom. This atom, as you know, is the basic atom because it has one electron and one proton. Every Bohr energy state for the atom is designated by an integer (non-fractional) called the principal quantum number of that energy state. An energy level integer of 2 is indicative of more energy than an energy state of 1, the ground energy level associated with the orbit of the moving electron cloud. When an atom is changed in state the difference in energy is in the form of radiation, either absorbed to increase the energy level of the atom, or emitted in order to permit a return of the atom to or toward its normal state. The energy of the quantum emitted is directly proportional to the frequency and inversely proportional to the wavelength. The wavelengths of the atom's radiation are stated in nanometers, 10^{-7}. centimeters in length. Figure 1-4 illustrates a basic energy-type diagram of atoms, and the wavelengths of the atoms expressed in angstroms. Note the classical names of the spectra series such as Lyman, Balmer, Paschen, Brackett, and Pfund. The Lyman series is associated with the lower frequencies or longer wavelengths, while the Pfund series is associated with the highest frequencies or shortest wavelengths.

THE DYE LASER

This type of laser emits light that exists only within a narrow range of wavelengths, but it can be tuned to operate over a relatively broad band. The lasing medium is a fluorescent dye that emits light rays or photons over a rather continuous spectrum when this material is pumped or energized by another type of laser. Usually an optical cavity is used to confine the emissions to a rather narrow band or even to a particular frequency. The pumping is done by a laser that emits an ultraviolet light and this, in turn, is guided into the cell containing the dye by means of a quartz lens. When considering the forces in the atoms as electromagnetic radiations, you can use the concepts of standing waves and single frequency generation and absorption by means of properly designed cavity resonators and still use some of the principles of optics such as beam splitters, birefringent crystal actions, mirrors, polarizing filters, diffraction gratings, and such to conduct experiments, make determinations, and create useful devices.

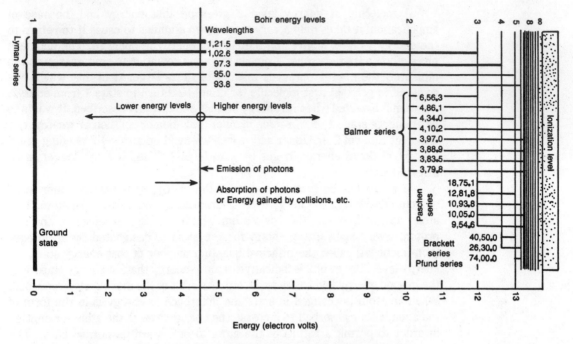

Fig. 1-4. A basic energy-level diagram (Bohr).

In the dye-type lasers, the molecules are certain kinds of dyes that normally have a tendency to absorb light photons easily and thus raise themselves to highly excited states. When a given wavelength is repeatedly passed through the dye laser being pumped by another laser of some kind, say a nitrogen laser (which is emitting ultraviolet radiations), the dye laser beam gains strength and is amplified tremendously. Normally, a diffraction grating is used to separate the desired frequency, and then a reflecting glass plate at the other end of the device causes the beam to pass back and forth through the dye many times.

CONSIDERING SOME LASER TERMS AND DEFINITIONS

As in all of science, one of the keys to understanding is to be able to determine what scientists and engineers are talking about. In mathematics, for example, if you don't know what a "transform" is, you are in trouble. So it is in the field of laser sciences. Here is some terminology commonly found in discussions of lasers.

Fabry-Perot Interferometer. Used in dye lasers to separate given wavelengths, it consists of two partially reflecting surfaces with a precise spacing through which radiations pass.

Etalon. A specific design of the Fabry-Perot interferometer made from a single glass plate. Its two parallel surfaces are coated with some partially reflecting materials. Light rays of different frequencies passing through this

16

device take different paths, and so some rays combine to reinforce themselves while others tend to destroy themselves by out-of-phase mixing. The phase is a function of the paths taken.

Spectroscopy. A means of causing the different wavelengths of light to separate due to traversal of some kind of medium and so produce a spectrum of light and dark (reinforcing and destructive) combinations of the frequencies. Most good physics texts discuss this concept in detail. There are many ways to accomplish this phenomena (Doppler spectroscopy, polarization of spectroscopy, photon spectroscopy and so on). A diagram to show the different wavelengths as related to atomic energy is shown in Fig. 1-5.

Gluons. An assumed carrier of a basic attractive force in an atom. It is called the strong force or the color force. It is useful in atomic theories and hypotheses and considered to be a massless particle.

Quarks. The assumed basic elements of protons, neutrons, nutrinos, and mesons. A quark is said to have the property of a collector charge and any such particle can then be said to be able to emit or attract a gluon. Quarks are theorized to be held together by gluons.

Glueball. The resultant of two gluons occurring when an exchange with other gluons occurs. A theoretical concept not experimentally confirmed as of this writing.

Neutron. A mass without charge found in the nucleus of the atom along with the protons that make up the positive charge. Normally, the charge between the protons and the electrons is balanced so the net charge of the atom is zero. If an electron is lost, the atoms become ionized or positively charged. If a proton is lost, the atom becomes negatively charged, assuming the electrons originally there remain in orbit (Bohr's concept) about the nucleus. Neutrons cannot be influenced by electrostatic or electromagnetic fields or magnetic or electric fields. They are said to possibly be influenced by a particle called a pi meson.

Pi Meson. A particle discovered in 1947 that is said to have a mass about 275 times that of an electron. It may interact with both the neutron and the proton at the core of the atom.

Photon. According to some theorists (including Yukawa) a constituent of the atom which has a rest mass of zero, as does the gluon. A photon is a particle that carries light energy. It is a part of the concept of light. Einstein assumed

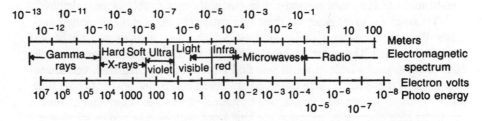

Fig. 1-5. Wavelengths and photon energy of the high-frequency radiations.

that the energy in a light beam travels through space in concentrated bunches called photons which energy is given by:

$$E = h\nu$$

where E is the energy in electron volts, h is 6.57×10^{-34} jouleseconds and ν (nu) is the frequency in cycles/second of cut-off, or stopping, of emitted electrons in a photoelectric effect. This is the cut-off frequency which is different for each emitting surface. The quantity $h\nu$ is the energy of a photon. A photon cannot emit or absorb another photon.

Color Force. A concept associated theoretically with he binding force of the atomic nucleus. It is currently theorized that the photon is possibly the carrier of an electromagnetic force that holds the atom together. But, in some current theories the gluon is thought to be the carrier of a so-called strong force or color force. It is then said that a gluon (associated with the atom-of-light analogue) has a color charge. (See Kenzo Ishikawa, Scientific American, Nov. 1982.)

Hadrons. A particle proposed by Murray Gell-Mann and George Sweig of CIT, which is subject to the strong force. In their theories, the fantasy of colors associated with forces predominates and flavors of colors are not unusual terms.

Up-Flavored Quark. An up-flavored quark is a part of a hadron and is designated by the symbol (u).

Down-Flavored Quark. Also a part of a hadron, designated by the symbol (d).

Strange-Flavored Quark. A particle which is theorized to explain certain mathematical long-lifetimes of some types of particles within an atomic force structure.

Anti-Quarks. Particles postulated to have opposite characteristics to the normal quark concept.

Quark Flavors. Names associated with two more concepts: the charm (c) and bottom (b).

From the above listing you can begin to get some idea of the strange manner in which theoretical physicists use various names of other well known phenomena or physically sensed qualities to assist in the explanation of the new modern theories of what makes up atoms. They constantly search for the real basic explanations to "What holds an atom together?" and "What are the constituents of the atom beyond those currently measurable or observable?"

To assist these magicians in their fabrication of (mathematical) models and other devices to explain their theories, which may or may not be of an advanced nature, the color concept is helpful in that you can diagram such objects as, say, glueballs in which you find gluons, and as theorized, you know that each of these has a force called color. These colors designate that each of these "things" have some unique mathematical properties. They do not mean the usual idea of color as we associate this word with blue or red or whatever, but it is easy for the mathematical physicists to be able to say "That gluon emits a blue gluon and so it becomes a red gluon." They assume that one gluon can emit many gluons and that the combinations of the total number in an atom make it colorless, or as one might think, it has some constant energy state although there is interaction inside this "world." The number of gluons in a glueball has not been suggested or well-defined in physics.

Fermion. Particles which have a spin (angular momentum) of 1/2 integer are called fermions. These are electrons, protons, neutrons, and quarks. It has been stated that all electrons have a characteristic "spin" — an angular momentum about some axis — and this has a value of:

$$\text{Electron Spin Momentum} = 0.52723 \times 10^{-34} \text{ joule-seconds}$$

Whether the particle actually spins, or whether this is just a loop of electric current (a spinning electric current-loop) is not known. What is known is that such spins give rise to magnetic moments or angular momenta. The alpha particle and the pions are said to have zero spin angular momentum.

In atomic and nuclear physics the elementary particles such as electrons, protons, and neutrons have an angular momentum associated with an intrinsic spinning motion, as well as motion about some external point. It is also recognized that the spin values can take on only definite discrete values rather than a whole series of smooth transitional values. Thus angular momentum as considered here is said to be quantized.

Employing the "right hand rule" of vector notation allows an elementary concept of a small particle with spin and angular momentum. (See Fig. 1-6.) The classical law of the conservation of angular momentum states that the sum of the external torques must be zero. This may or may not be considered by advanced theories in particle dynamics of the atoms.

THE FUNDAMENTAL THEORIES

The fundamental theory of electrodynamic interaction of charged particles in an atom is called quantum electrodynamics. In this concept, the force between two electrically charged particles can be accounted for by an exchange of photons (the photon itself being electrically neutral). This theory does not account for the change in charge which may exist from a particle emitting a photon.

The basic theory of a color force may help. (Remember, colors as referred to in this concept and in the modern atomic theory are not really colors at all.) Color means a mathematical statement of some kind and complexity, which has some significance with regard to size, mass, charge, type and so on of the particles. By manipulating the mathematics they say they can illustrate changes of such objects as gluons, from a red color to a blue color.

In a way this is a nice idea to describe what might be absolutely confusing (advanced mathematics) in terms that are only a little confusing! Laser light is referred to as red or green or blue or violet because it can be perceived as having this frequency of oscillation, which our eyes, nerves, and brain then

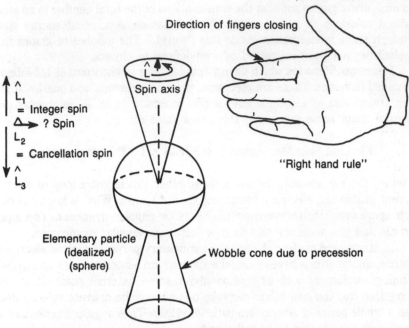

Fig. 1-6. Idealized concept of a particle with spin and angular motion.

interpret as a color. You must know when color statements are associated with which particular theory of particles, of objects, or light beams. You must also have some idea of what the word color means in that language frame. In a way it is like the American words bear (the animal) and bear (to carry) which sounds alike, are spelled alike, but whose meaning depends upon how they are used in what context.

The atom is not as simple as you are often lead to believe. In atomic physics as, the various particles and forces are discovered and better understood, theories of far-reaching importance may be forthcoming. For example, laser beams are rapidly attenuated by atmospheric molecules. Research may find a way to overcome this problem!

MORE DEFINITIONS AND CONCEPTS OF MODERN ATOMIC PHYSICS

Let's expand the definition of a quark and abbreviate an up quark as u. The down quark is designated as d. These states are called flavor, which sounds like a tongue-in-cheek statement by some very advanced people. In this concept, the proton is made up of uud and a neutron is made up of udd. It has been a third kind of quark, which has strange properties (long lifetimes compared to others) is designated s. With these kinds of designations, scientists become very happy for they can describe various kinds of elements or particles in this kind of symbology. It becomes a kind of short hand way to write things and then manipulate the written abbreviations in a mathematical methodology.

The idea of color was first proposed to make the quark model jibe with the Pauli Exclusion Principle. This principle says that any two fermions cannot share the same quantum mechanical state, which means they cannot have the same energy value, spin value, or any other numbers which are used to identify one fermion.

The fundamental theory of the electromagnetic interactions of particles is called quantum electrodynamics (abbreviated QED). It used to mean "The statement, hypothesis, or whatever is now considered to be proven." If you are a modern physicist, it means quantum electrodynamics. You may find it enjoyable to learn the new terms and abbreviations and definitions of science — some are very peculiar!

Quantum chromodynamics is the basic theory of the color force. It is abbreviated QCD. The mathematical framework for this theory was developed in 1954 by C.N. Yang of the State University of New York and Robert Mills of Ohio State University. QCD states that one colored particle interacts with another by the exchange of gluons. It is said that because there are three colors that predominate, the theory of QCD is much more complicated than that of QED. QED is the theory most of us are familiar with. The gluon colors are blue, red, and green. Sound like television? But remember that these aren't colors at all as we know them. They are mathematical expressions or statements. Among the elementary particles which have been found in an atom are Leptons, six of which are known. One of these is called the electron, another is called a muon, and a third is called a tau. These are like electrons except for mass. The muon is said to be 200 times as massive as an electron, and the tau is said to be 3,500 times more massive than the electron.

ON THE ANGULAR MOMENTUM OF LIGHT

We have discussed the angular momentum of various constituents of an atom. Now let's consider the idea of momentum as associated with light rays or photons. Light rays can deliver linear momentum to an absorbing screen or to a mirror in accord with the classical theory of electromagnetism. Also, light rays can be polarized. Thus, considering circular polarization, it is not difficult to imagine that angular momentum might somehow be associated with it. Circular polarization is considered as some kind of rotation of the $\hat{E}$ vector of the electromagnetic representation. The proof of this concept was shown by Beth in 1936 when in his experiments of producing circularly polarized light in a doubly refracting slab, that slab experienced a reaction torque.

So it was hypothesized that if light photons carry away some angular momentum energy as they leave an atom, the remaining angular momentum of the atom must be reduced by exactly that amount carried away. Otherwise the conservation of angular momentum law is not satisfied.

There has been developed an expression, a simple one, which says that if a beam of circularly polarized light is completely absorbed by the object upon which it falls, an angular momentum:

$$L = \frac{U}{\omega}$$

is transferred to that absorbing object. U is momentum and ω is the angular frequency of the light rays. Thus we find that if L is the total angular momentum and U is constant, then the smaller the radian frequency ω the larger the transfer of angular momentum energy.

In some years past many studies (probably still valid and still being considered) were made of space ships with sun sails. These would be large areas of absorbent materials which would absorb sunlight and thus impart to the space ship a linear momentum, which, in turn, becomes movement through space. It may yet come to pass that some space ships will use some kind of sun sails to traverse the distant cosmic seas.

It is also interesting excitation of the atom causes the production of laser light rays. There may be unknown side-effects caused by stimulation of the other constituents of the atom may or may not be known as of today. If not known, then when they are determined, we may find some startling new concepts that are above and beyond the concept of just producing laser light.

Chapter 2

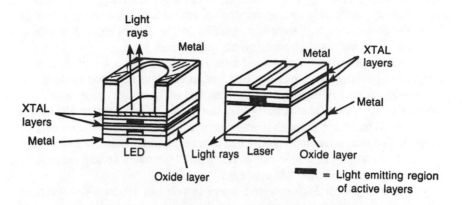

Some Basic Types of Lasers

IN SOME TYPES OF LASERS A CAVITY RESONATES AT LIGHT FREQUENCIES. THIS cavity is like a tuned circuit in its operation. (Good books on radar will enhance this explanation.) Let's discuss the transmittance of energy inside and through such devices. When light frequencies are desired, mirrors are used to cause reflections in laser cavities so energy is imparted to solid-state elements (rubies), gases (helium) or dioxide mixtures, and dyes, which absorb and release energy in the form of light photons at certain frequencies. The cavity, which is a tuned circuit, resonates only at one frequency with a relatively narrow bandwidth. This enhances the creation and transmission of a specific laser frequency of light which is monochromatic and has a single polarization.

LIGHT-REFLECTING MIRRORS FOR LASER BEAMS

Some mirrors are designed to absorb certain light wavelengths and pass other wavelengths. Thus, they may act as filters. For example, some mirrors reflect heat (infrared) and pass visible light. Others reflect visible light and pass infrared. These latter types may look like metallic mirrors, but may actually be coatings on highly reflective heat-resistant pyrex glass. Mirrors used with lasers may be called reflectors and may be polished on only one side to a very high degree of flatness and surface polish. They may be uncoated substrates with a coating of silicon monoxide or aluminum and the flatness may vary by as little as 1/20 of one wavelength.

There is a mirror called a front surface concave reflector. It is ground and

23

polished to a highly spherical surface and has the coatings just described on a type of substrate. The concave side of the mirror is highly polished, while the back side is flat. The edges are bevelled. It has a given focal length, concentrating the rays impinging on the concave surface back to a point some number of wavelengths away, and somewhat parallel to the arrival axis. The size is from two to six inches depending on the application.

The laser cavity end-mirrors and interferometer mirrors are made from a substrate of polished glass or silica that has appropriate reflector coating. On the outside, an anti-reflective coating is applied so laser light rays won't get back inside the cavity. The latter type of mirror is a window which allows the laser beam to burst out through this mirror when it reaches a given density or strength. The mirror surface on the inside is required so that the laser beam bounces back and forth inside the cavity until it becomes strong enough to emerge. Figure 2-1 shows this concept.

Some mirrors for high-powered lasers (which can produce ten watts or more of energy output) are concave or flat and have a coating surface of gold-copper or hard gold. The base is occluded nickel coated on copper. This material has a very high heat resistance and a very high thermal conductivity. The mirrors are total reflectors and can be used up to the multimegawatt region of power outputs.

Gold is also used in another way, by coating a pyrex substrate. Oriel Co. does this and produces a mirror for low-powered laser systems. These mirrors are polished to within 1/80 wavelength and the gold coating gives a high degree of reflectivity in the infrared region.

This brings up the concept of roughness. You usually think of polishing something as simply removing black splotches or some undesired coloration. In a woodworking sense, you can think of smoothing down a wood surface until you can feel no grain by hand. From out into space, the earth's surface looks

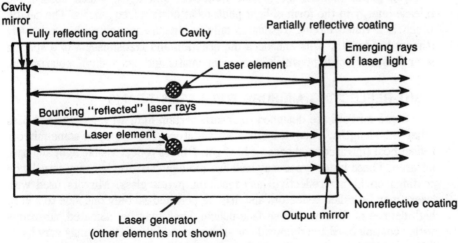

Fig. 2-1. Mirrors are used in a laser cavity to bounce rays back and forth to stimulate emission.

absolutely smooth, although you know it is not because you live here. When you look at a surface under some microscopic magnification, you'll find that it too is not perfectly smooth. Did you know the earth's surface is as smooth as a bowling ball? It is, and that tells us something about the surface of a bowling ball. The next time you run your hand over such a surface, think about the relative smoothness. This concept is extendable to the smoothness of laser mirrors, which may be polished to within 1/80 of a wavelength of light.

An angstrom is one ten billionth of a meter, and light is around 6400 of these units. When you consider a surface being polished to 1/80 the wavelength of light, you are looking at a surface with hills and valleys which deviate no more than 800 one-ten billionths (.000000800) of a meter!

Zinc selenide is used for a reflecting mirror in some lasers. It has a low absorption of the light rays—around 10.6 microns. Germanium is also used, but germanium has a thermal runaway tendency. It has a relatively high thermal conductivity. The absorption of the energy by germanium increases with increasing temperatures, so with this type coating it is important to keep the temperature below 40 degrees centigrade. An uncoated germanium reflector reflects about 36 percent of the impinging energy and that is good.

MORE ON LENSES AND FOCUSING OF LASER BEAMS

It is of primary importance when using laser beams that they are focused to extract energy to do any kind of work that involves temperatures. The next chapter explains this in more detail. At this point, let's examine some lenses and focusing concepts as applicable to beam-types of light; light-control once that light has been generated.

Contrary to some ideas, some laser beams are not monochromatic but are composed of many different light frequencies near the very strong center of resonance frequency. This is probably due to some randomness in the stimulation-relaxation action of atoms that may not move precisely together. Dye lasers may have a broad-frequency beam. Other types may resonate in such a manner in generation and have such filtering that only a very narrow band of frequencies is produced.

When you think of focusing a laser beam, consider lenses that do not discriminate among the various light frequencies as much as a filter-system might. Of course, the propagation of various frequencies through a dielectric medium that is transparent to those frequencies offers more delay to some frequencies than to others. Only certain light frequencies are focused precisely.

Indulging in some speculation, however, consider the light from the sun to be nonmonochromatic and therefore composed of many frequencies. Yet you know that we can focus some of these frequencies with an ordinary magnifying glass, concentrating the energy in them to such an extent as to burn holes in wood or paper. In this sense, the focusing system is independant of light frequency. The focusing system is adjusted to a given band of light frequencies when it is focused.

Light waves propagate, like radio, radar, or microwaves, in a direction perpendicular (normal) to the wave front. Light travels at different speeds in different materials and the ratio of the velocity of light in a vacuum to that in a given type of material is known as the index of refraction.

Many people have trouble with this concept of changing speed of light. It may not make sense to say that at some times the speed of light is slower or faster than at other times. Everything we know is based on a constant speed of light, some 300,000,000 meters/second. The resolution of the problem can be found in a sage's statement: "The speed of light is constant over an infinite distance. Therefore, one can find variations in the speed of light along the way. I tried to imagine a thread that extended to infinity and to imagine a kink in that thread. When viewed from an infinite distance the kink did not seem to exist at all but it was still there. That kink could be interpreted as being the change in velocity during the infinitesimal part of a second during which time that ray might pass through a different medium." This philosophy, explains radar wave-guides and laser cavities.

The basis for lenses is Snell's law, which says that as light moves through one medium into another it is bent according to:

$$N_1 \sin \theta = N_2 \sin \phi$$

If you draw a line which is perpendicular (normal) to the side view of a lens, the angle θ is the angle between the direction the light ray takes and that normal-line. This is called the angle of incidence, or entrance angle into the lens. The outgoing line of the ray is at another angle called phi prime ϕ, the angle of refraction (bending) between the emitted ray and the normal-line through the lens. N is the index of refraction of the material for the incident ray, and N prime N' is the index of refraction of the material of the lens. Essentially what we mean is that the first N is the index of refraction of light in air approaching the lens and N' is the index of refraction of the changing medium of the glass lens. Light bends toward the normal-line in the higher index of refractive materials; in this case, air. It bends away from the normal-line in the lower of the two refractive materials; in this case, the lens.

When an optical engineer begins analyzing the ray progression through various lenses he makes some assumptions. First, he assumes that the angles are small and so the sine value can be replaced by the angle in radians. (This is a standard mathematical approximation when the angle is 10^0 or less. The sine $0^0 = 0^0$ when measured in radians.) Second, he assumes that the lenses are thin. Be aware that there is a whole engineering course that considers thin lenses and thick lenses. They are not the same. Trigonometry and geometry can be coupled with Snell's Law to produce a tracing of the rays through thin lenses with fairly good accuracy and relative ease.

Now, let's consider a thin lens and an object or light source that is located at some distance (x) to the left of the lens. The light rays from this object or source are incident on the lens through a distance (y). Examine Fig. 2-2. If an object is located off the principal axis of the lens and light rays reflected from the object pass through the lens as shown in Fig. 2-2, then magnification has

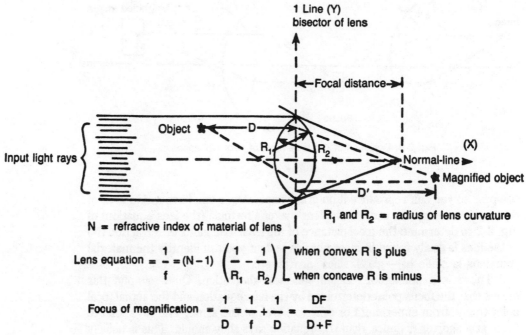

Input light rays

1 Line (Y)
bisector of lens

←—Focal distance—→

Object

R₁ R₂

Normal-line ▸ (X)

★ Magnified object

D′

R₁ and R₂ = radius of lens curvature

N = refractive index of material of lens

Lens equation = $\dfrac{1}{f} = (N-1)\left(\dfrac{1}{R_1} - \dfrac{1}{R_2}\right)$ $\left[\begin{array}{l}\text{when convex R is plus} \\ \text{when concave R is minus}\end{array}\right]$

Focus of magnification = $\dfrac{1}{D} = \dfrac{1}{f} + \dfrac{1}{D} = \dfrac{DF}{D+F}$

Fig. 2-2. Focus and magnification of an image with a thin lens.

taken place if the image is larger than the object. This can be expressed mathematically as:

$$\text{Magnification} = M\dfrac{D'}{D}$$

This equation says the ratio of focusing distances on each side of the lens govern the magnification factor. Find D and D′ on the diagram and note the difference in distances. You should be able to measure in centimeters the different distances and thus gain some idea of the amount of magnification of the lens.

There are situations where more than one lens is used to focus. In these cases, use the focused image of lens number one as the input to lens number two and cause the image to be at the proper position to focus into lens number two. There is also an equation for two thin lenses:

$$\dfrac{1}{f} = \dfrac{1}{f_1} + \dfrac{1}{2f_2} - \dfrac{2S}{f_1 f_2}$$

where S is the separation between lenses. Figure 2-3 illustrates this situation. The details of curvature extended beyond just the lens outline. A compass is used for this purpose. If you can get the curvature of any lens, open your

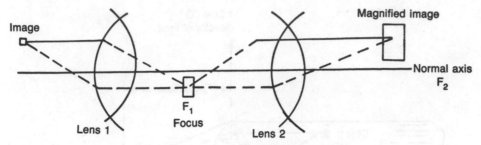

Fig. 2-3. Definition of the focal point of two thin lenses.

compass so you can reproduce it on graph paper. Once you have a side view of the unknown lens, you can use the lens-maker's formula (the lens equation) of Fig. 2-2 to determine the focal distance of the lens. The index of refraction (N) of the lens is easily found in most physics texts if you can identify the material your lens is made of — glass, silica, quartz, etc.

There is a phenomena called spherical aberration. Quite simply, this means that the focus point determined by the lens equation and the actual focal point (found from experiment or practice) are not the same. The rays tend to focus at a shorter distance than the equation says they should. This is usually true with thick lenses. The abbreviation for this phenomena is SA.

Let's discuss a few types of lenses, some of which are called plano lens types. They are flat on one side and curved (concave or convex) on the other. The spherical aberration is different for the different types of lenses. On a plano lens, for example, the SA is minimum if the flat side points toward the focal point. This means that you place the lens so the curved side faces the object or light source to be passed through the lens. When using a lens of the type shown in Figs. 2-2 and 2-3 (double convex types) the SA is minimum when the focal distances D and D' are equal. Finally, if you use two plano lenses placed so that the curved sides face each other, you can get a magnification with minimum SA. In this situation, you want all the light rays to emerge from the curved side of the first lens in a parallel manner and enter the curved side of the second lens and depart the second lens's flat side to converge and thus focus to a point. Of course, the image object that produces the initial light rays into the first lens's flat side must be positioned a distance away from that lens so that you have this image or object at the first lens's focal point.

Some lenses are not smooth curves, but are non-smooth curves over their entire face. These consist of curved parts or portions and straight line curves or portions all joined together. The lens is cut this way to minimize its sphericity. Such a lens is aspheric (nonspherical) in shape. The reason for making such a lens is to reduce the focal distance and minimize the SA.

In camera lens systems there are many lenses of different types, plano, double convex, convex-concave, and double concave. It is not uncommon to find seven or more lenses grouped together in a camera lens mount. That is why they are costly. At least one reason for doing this is to minimize SA and

28

provide good focus at proper distances for the camera body. Each lens must focus into the next in a proper manner. A study of camera lens systems in other texts is recommended if you want to get into the subject further.

Chromatic aberration is concerned with the color or wavelengths of light. A monochromatic system is a one frequency system or uses one color of light. If you have a lens which tends to separate the focal distances of various colors or wavelengths of light, due to different speeds of travel through the lens for these different components, then you have chromatic aberration. There are lenses designed so the wavelengths of light, over a given band, all travel the same speed through the lens and focus at the same outside point. These are called achromatic lenses.

Now let's examine a concept of the use of lenses in a system that uses laser light beams. One beam is to excite or pump-up a dye laser and the second beam emits from the dye cell laser device. Notice the use of lenses for focusing the output beam to provide a rather wide beam, which then can be filtered to produce a monochromatic light ray of the right frequency to excite the dye cell's atoms. (See Fig. 2-4.) Lens 1 and lens 2 comprise a kind of telescope arrangement, and they act in a reciprocal (two direction) arrangement. Lens 3 is simply a focusing double-convex type of lens used to focus the input excitation beam into a smaller more concentrated beam to excite the dye-cell atoms.

Of course, there are other types of lenses and different shapes of lenses. The purpose of any lens is to provide a focal point for input and output. Lenses are necessary with laser beams to concentrate the energy precisely where it should be for the purpose it is to be used. Do not think that because a laser beam is not focused that it may not blind you. It might! Always be careful when experimenting with any kind of focused light or laser beam. Don't look directly into it—EVER!

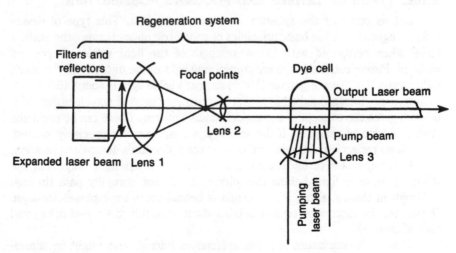

Fig. 2-4. A lens arrangement for a dye laser.

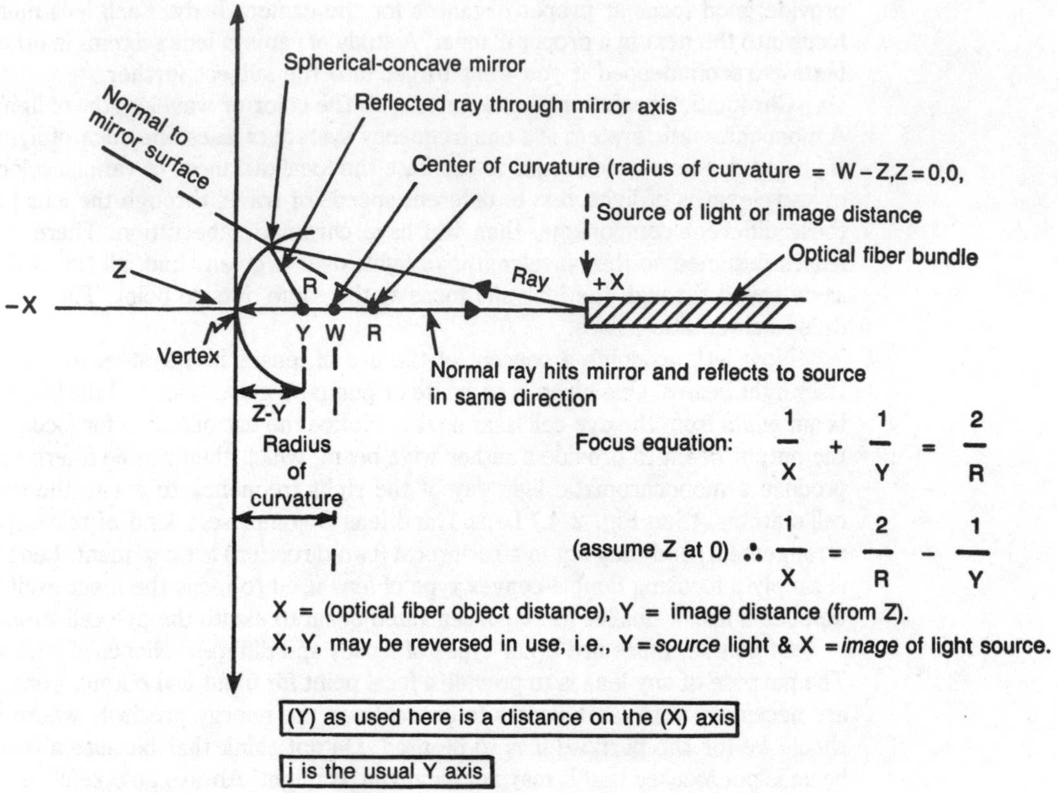

Spherical-concave mirror
Normal to mirror surface
Reflected ray through mirror axis
Center of curvature (radius of curvature = W−Z,Z=0,0,
Source of light or image distance
Optical fiber bundle
Z
R
Ray
+X
−X
Y W R
Vertex
Normal ray hits mirror and reflects to source in same direction
Z-Y
Radius of curvature
Focus equation: 1/X + 1/Y = 2/R
(assume Z at 0) ∴ 1/X = 2/R − 1/Y
X = (optical fiber object distance), Y = image distance (from Z).
X, Y, may be reversed in use, i.e., Y = source light & X = image of light source.
(Y) as used here is a distance on the (X) axis
j is the usual Y axis

Fig. 2-5. Details of a spherically concave mirror. (Note that "y" is actually a distance along the "x" axis.)

SOME TYPES OF LENSES AND FOCUSING POSSIBILITIES

Let us consider the spherical concave mirror first. This type of device reflects light rays if the basic principles of geometric optics (having the surface large when compared with the wavelength of the light reflected) are not violated. Please recall that we are considering light in the micron (10^{-6} meter), millimicron (10^{-9}), or angstrom (10^{-10}) region. Visible light is about 450 to 675 millimicrons. Infrared light has higher frequencies (shorter wavelengths). Figure 2-5 gives the concept of a spherical concave mirror. There can be problems with this type of reflector. If the source and image are not precisely located, there is no reflection of the source seen at the reflection's designated position.

It is important to understand that a mirror reflects light only from the front. If there is light behind the mirror, it will not normally pass through although in theory, a ray whose origin is behind the mirror passes through. (Later on, the discussion on lenses talks about such things a a real image and virtual image.)

Due to the curvature of a concave/convex mirror, rays might be considered to converge/diverge. If they converge, there is a concentration of the light rays. A simple statement of geometry [image (i) = 1/2 r (radius of curvature) = F (focal distance)] tells us where the focal point of such a mirror should be located. It is at 1/2 the radius of curvature distance from the vertex. The classic equation expression for the focal distance is:

$$\frac{1}{0} + \frac{1}{i} = \frac{1}{F}$$

where 0 = source distance (object distance), i = image distance, and F = focal length (positive if image is real, negative if image is virtual).

PLANE MIRRORS

Light which impacts on a plane mirror is reflected from that mirror through the angle of incidence. To put it another way, the angle of reflection is equal to the angle of incidence. This is easy to check out in a simple experiment. Figure 2-6 might help. The plane mirror is important in some types of lasers. In a ruby rod laser, the stimulated light energy bounces back and forth until it emits from one end of the rod. The other end is a polished, opaque, plane-mirror surface.

When you view the image from a plane mirror, be careful to observe the interchange of left and right. If you think about it for a moment you'll realize that a man who shaves in the mornings shaves with his left hand on the left side

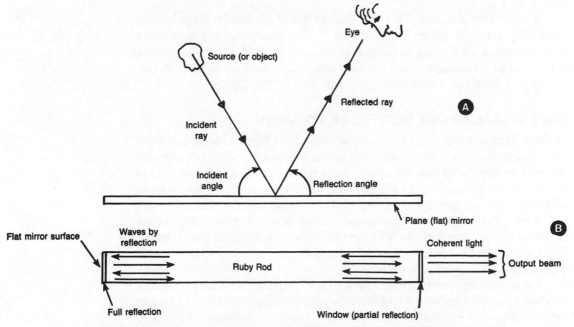

Fig. 2-6. Reflections from a plane mirror. A flat mirror plate reflects light images (A). A ruby-rod in a laser must reflect light to stimulate emission of a laser beam (B).

31

of his body if he is right handed and vice versa. Our minds adjust to this phenomena and we usually aren't even aware of it, but think about it next time you face yourself in the plane mirror. Remember a plane mirror reverses right and left. This may or may not be of consequence in reflection of light rays and light energy.

THE FOCUSING PROBLEM

Let's consider getting light into an optical fiber (singly or a bunch of fibers all terminated in a flat and polished end) by using a reflecting device such as a mirror or use a focusing device such as a lens. Take into consideration the angle of entry into the fibers of the light energy. If the light energy is presented through too wide an angle, the angle of reflection from the cladding may exceed the critical or Brewster angle and hence might not propagate through the fiber. We want to concentrate the energy and light rays within a cone which has the requisite acceptance angle for the mode of propagation we desire. Size becomes important, both the physical size of the lens or reflector and the relative physical size of the optical fiber cable or strand. If the geometry of the reflector or lens can be made compatible with requirements for reflections and focusing and is small enough so that the cone angle is within tolerance, then you have it made, but sometimes that takes some doing. The small size (diameter) of the optical fibers makes fabrication of such lenses and reflectors somewhat difficult. Credit is due to the wonderful manufacturers and their research departments. They have been able to accomplish a formidable task! (See Fig. 2-7.)

If the mirror is small and of a size close to that of the fiber, it not only loses much of its ability to gather light rays, but it can present what it gets within a relatively small cone angle to the end of the optical fiber. The larger mirror gathers more light from a source but the angle to the fiber may be much larger. Ideally, a source has a lens about the same size as the fiber.

DOES A LASER BEAM NEED TO BE FOCUSED?

It is normally assumed that you know that a laser beam is focused. It is the focused diameter of the beam and the direction of travel of the beam when used in a cutting application that determines the kerf (the parallelism of the sides of the cut in the workpiece). As you can easily imagine, one might have a beam which comes down into the metal, in a kind of slanting direction giving rise to a "V" type of cut (broad at the top and narrow at the bottom). This is not desirable in many applications. With proper focusing and cutting direction, the kerf (or cut) can have parallel sides. They might be from narrow to rather wide depending on the laser's dimensions.

Let's examine some concepts of the laser beam. First of all, recall your childhood experiences with a magnifying glass. To make something burn, you held the glass so it focused the sun's rays into a very small dot on the object to be ignited. The smaller the dot, the hotter the point became. Soon it smouldered, and in some cases burst into flame.

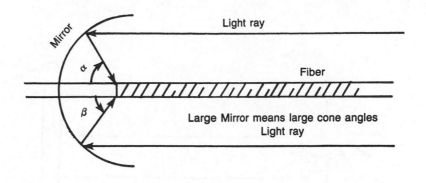

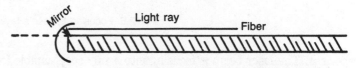

Large Mirror means large cone angles
Light ray

Smaller reflector may get less source light
but angle of reflection may be smaller

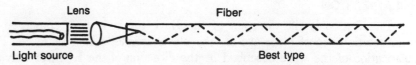

Fig. 2-7. Presenting light rays to a fiber within an acceptance cone angle.

A laser beam is normally focused through a special lens so its concentrated energy is directed into a tiny spot like the sun's rays. The tremendous amount of energy concentrated there can burn through metal, vaporize it, or weld it. That focal distance is very sharply defined. For example, if the spot is two inches away from the beam outlet on the laser head, then just microns of movement farther or nearer reduces the energy to the spot. This becomes important in eye surgery, where a laser beam can be focused through the eye to the back wall and there used to tack or weld the retina in place. The beam does not damage the eye as it passes through it because it is not focused except at the retina. Figure 2-8 illustrates the concept of lens focusing. Of course, there is more to the story than stated here. Polarization of the laser beam (Fig. 2-9) plays an important part in the focusing concept, and polarization can be controlled or changed with proper quarter-wave reflective phase units.

After the laser beam starts cutting, the metal absorbs the energy from the beam in a parallel-incident manner to the impinging rays and the metal melts or

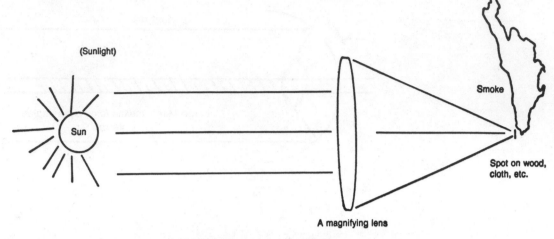

(Sunlight)

Sun

Smoke

Spot on wood,
cloth, etc.

A magnifying lens

Fig. 2-8. Focusing the sun's rays.

vaporizes. The laser beam is focusing into a tiny spot suitable for the task at hand. The energy it contains heats, melts, vaporizes, or cauterizes the object upon which it falls. If less energy is used, either by using less laser power or pulsing the laser less frequently, the energy applied is less and can thus produce other effects which are useful.

POLARIZATION

Polarization is the direction of the $\hat{E}$ vector in a wave composed of electrostatic lines of force and $\hat{H}$, the electromagnetic lines of force. The polarization of the wave described by the direction of the $\hat{E}$ vector may be varied. The polarization may be parallel, circular, or perpendicular to the direction of the light ray. Faraday rotation of an electromagnetic wave means that upon impact with a reflecting material, the $\hat{E}$ and $\hat{H}$ vectors rotate in opposite directions to each other and to the direction they had upon incidence with the material.

When describing the phenomena of refraction, reflection, and transmissibility of laser rays through or from surfaces of various materials, it is customary to indicate what happens to the polarization of the ray as it passes through or is refracted or reflected from such a surface. Let's assume polarization is related to phase. Since you know how phase combinations affect the amplitude of a resultant wave and the phase of a resultant wave, you have some idea of what happens in laser light rays if such rays are divided in any manner and then recombined. For example, let's use a beam splitter that accurately divides the power in a laser beam exactly in half. Suppose that the polarization between the two resultant beams is changed by 180 degrees and then the beams are recombined. Intuitively you would say that the two beams cease to exist since they cancel. This is the same idea used with phase relationships in any type of electrical or electronic circuit, radio, radar, or microwaves.

34

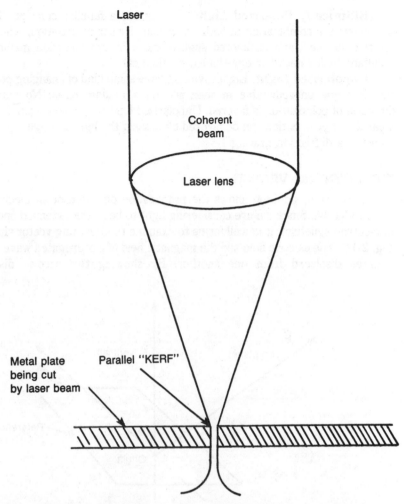

Fig. 2-9. The focus concept applied to a laser beam used in cutting metal.

The subject of light rays and their use and control leads quickly into the subject of polarization. Some definitions are needed!

Linearly Polarized Light. The condition where the electric e vector, describing the electric portion of the wave, vibrates in one direction only, and that direction is perpendicular to the direction of propagation of the wave.

Circularly Polarized Light. The direction of vibration of the wave rotates around the direction of propagation in a corkscrew fashion. No single direction of vibration is favored. Be aware that the direction of rotation of a circularly polarized wave can be in either direction—right-handed or left-handed. This refers to the concept of vectors where, if the fingers of the hand curl in the direction of rotation, the thumb points in the direction of propagation.

Elliptically Polarized Light. This is not a familiar concept. This is somewhat the combination of both linear and circular polarization, where the rotation of the wave is favored slightly more in one direction making the resultant look somewhat egg-shaped or elliptical.

Unpolarized Light. Light rays that have some kind of changing polarization in some unpredictable manner and on a random basis. No particular direction of polarization is favored. Unpolarized light can be segregated so that a particular polarization can be favored by passing the light through a polarizer or using a diffraction grating.

THE POYNTING VECTOR

This vector addition shows the propagation direction of an electromagnetic radiation. Since we are considering light to be in the extended spectrum of electromagnetism, it is well for us to examine the Poynting vector shown in Fig. 2-10. The electric field and the magnetic field of a propagated wave are 90 degrees displaced from one another. By showing two arrows displaced

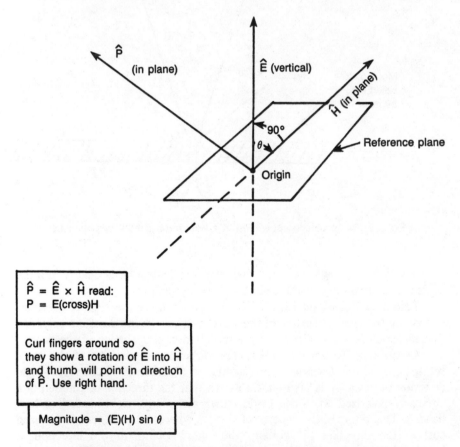

$\hat{P} = \hat{E} \times \hat{H}$ read:
P = E(cross)H

Curl fingers around so they show a rotation of $\hat{E}$ into $\hat{H}$ and thumb will point in direction of $\hat{P}$. Use right hand.

Magnitude = (E)(H) sin θ

Fig. 2-10. Method of vector multiplication on the poynting vector.

36

by 90 degrees, whose length represents the instantaneous direction of those waves, the vector addition of the two represents the total or the Poynting vector. You "curl" the $\hat{E}$ vector into the $\hat{H}$ vector and the thumb points in the direction of the resultant $\hat{P}$ vector. The magnitude of the resultant is the absolute value numerically of E and H multiplied together times the sine of the angle between them. When this angle is 90 degrees, the sine is 1.

The above demonstrates how a light beam P might be decomposed into its two orthogonal (right angle) elements, E and H, whose amplitudes are such that with vector addition they can be recombined into the original beam. It is sometimes said that a beam of unpolarized light can be represented as two right-angle components of linearly polarized light which are equal in amplitude.

THE EFFECT OF LIGHT POLARIZERS

These devices tend to pass light components that are polarized in the same manner as the polarizer. They can be made to act as light switches. Two of them they are like sunglasses that pass only a given polarization. If you rotate one of them so that its direction of polarization is 90 degrees away from the first, then no light is passed. In a sense, you have a light switch, and since a laser beam is monochromatic and is polarized light, you can use such a switch very effectively in some applications where laser beam control is useful.

But nothing is perfect. There is not a perfect extinction of the light transmission even when the polarizers are properly adjusted. The rate of light transmission when the polarizers pass the light, to the rate of light transmission when the polarizers are adjusted to cut-off the light transmission, has a limit. It is called the extinction ratio of the polarizers.

METHODS OF POLARIZING LIGHT

The laser generates the in-phase or polarized light ray. Light can also be polarized by reflection at other than some normal incidence from an interface of two dielectric materials of different indices of refraction, such as that between air and glass. If the angle the light hits the reflecting surface is equal to the Brewster angle (equal to that angle whose tangent is $\frac{n1}{n2}$) then the reflected component of the beam is entirely polarized perpendicular to the plane of incidence. The part of the beam that penetrates the interface is small and is partly polarized parallel to the incidence plane. By considering two materials which have different indices of refraction, and passing a light through them at a proper angle, you get a perpendicularly polarized component of the light beam or a parallel component of the light beam. This also is useful in some applications.

You can control the flow of a light beam by having it polarized and using polarizers in a proper manner. The same idea is true with electromagnetic radiations in radar and microwave applications. Ferromagnetic devices can control the polarization of radar and microwaves in much the same manner that some types of crystals, glass, and other substances control the polarization in light beams. There is a device known as a wire grid that can polarize the longer wavelengths in the infrared spectrum, if the spacing of the grid is small compared to the wavelength.

It is possible to absorb one component of a light beam that has some given polarization. This is called polarization by dichroism. The phenomena is accomplished by orienting a long chain of molecules in one direction so light that is vibrating parallel to them will pass through the molecular material. Oriel Corp. makes sheet polarizers made of an absorption of iodine in a sheet of thin polyvinyl alcohol that has been stretched so it causes the molecules to be stretched in a long chain. It is a polarizer for light beams of given frequencies.

To further understand the control of light rays, refer to the calcite crystal. This is a parallelepiped type of crystal with a well defined optic axis as illustrated in Fig. 2-11. As you note from the figure, there are many parallels to the optic axis, which is a direction through the crystal. This crystal is peculiar, for it shows a double refraction (two beams emerging from the crystal) when the incident beam is perpendicular to the optic axis. If the incident beam's direction is parallel to the optic axis, there is no double refraction.

Now we come to another definition which is important in optics and light control. One part of the beam which passes through the calcite crystal is called the ordinary or O ray. The second component of the incident beam we call the extraordinary ray or E ray. The polarization of the ordinary ray is parallel to the direction of travel. The extraordinary ray is polarized perpendicular to the

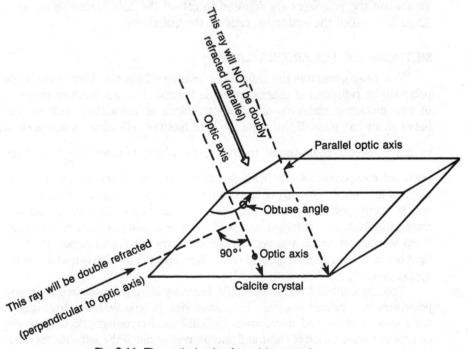

Fig. 2-11. The optical axis of a calcite crystal.

direction the beam is traveling. Thus the polarization of the ray component of the beam determines whether they ray is called an ordinary ray or an extraordinary ray, and has the abbreviation O or E.

So, what's the significance of this? When a beam having these two polarized components (ordinary light has them) passes through a calcite crystal, the beam is essentially split because the index of refraction of the crystal varies with the different polarizations of the two rays. The index of refraction varies from 1.66 for an E ray to 1.49 depending upon whether that E ray component travels parallel to the optic axis or perpendicular to it. By shifting the physical orientation of such a crystal with respect to a beam of laser light, the refraction of that component of the beam can be shifted as the index of refraction is shifted.

If an unpolarized beam is passed through a calcite crystal, the light becomes polarized and has two components perpendicular to the crystal's optic axis. One is the O ray and the other the familiar E ray. Now here is a fascination part of this study. Both rays travel through the crystal at the same speed but they are refracted differently, so the difference becomes evident. The O ray normally is refracted according to Snell's law of optics, but the E ray changes its direction somewhat, due to the effect of a different index of refraction. It moves away from the optic axis and out of the plane of incidence. The emergence of the two rays from the crystal are parallel to the incident beam direction, but both are shifted somewhat due to the effect of the crystal. They now emerge in parallel lines. In effect, the O ray separates from the E ray, and two light beams emerge where there was just one incident beam. As stated earlier, when a calcite crystal is placed on a page of a book like this, the lines and letters appear doubled. To state it another way, you see two sets of lines and characters. This phenomena is called double refraction and has its place and usefulness in handling and controlling light rays. A final thought: Some scientists believe that the speed of travel of the two rays may be different, and if the index of refraction is defined as $\frac{c}{v}$, where c is the velocity of light and v is the actual beam speed, there is a variance in the index of refraction for one ray with respect to the other, so more bending of one ray than the other takes place.

There are several types of doubly refracting crystals such as ice, quartz, siderite, calcite, dolmite, and wurzite, to name a few. Each exhibits this ray splitting phenomena when a beam of light at about 5890 angstroms is used as the incident beam. The frequency of the incident beam may be of importance. What may be a phenomena and effect with a beam of one frequency and wavelength may not appear with a different light beam of a different wavelength and frequency. Also, this assumes a nonpolarized incident beam. Just what the effect of passing various frequency laser beams through these crystals is, I leave to your scientific curiosity and experimentation. The laser beams, remember, are polarized. Thus the orientation of the crystal with respect to a laser beam produces various effects.

Still concerned with polarization, let's examine the types of prisms which use the polarization effect of the light rays. A polarizing prism is as good as its name: It polarizes the ray passing through it. To polarize some incident light, you can use a polarizing prism to do the job. It passes the E ray, restricts passage of the O ray, and the result is a beam of polarized light. One type of prism is made of two wedges positioned so that there is a tiny air space between them (Fig. 2-12).

If you focus some unpolarized light on the element so that it enters the crystal prism perpendicular to the optic axis, the (E) ray goes through the first prism, hits the air space and then goes through the second correcting prism and emerges essentially unchanged. The O ray takes off from the prism-air interface at some wide angle and gets lost somewhere else. Thus the two components have been separated, and just the polarized (E) beam is left coming through the element which is what was wanted in the first place. The prisms can be made from calcite.

There is a critical acceptance entry angle for the incident beam. If too low an angle is used, it is reflected, if too great an angle is used, some O ray gets into the desired output and messes up our linearly-polarized E ray. The acceptance angle for some prisms of this type is around 9 degrees. Oriel Co. makes such a prism polarizer with extinction ratios of up to 10^5 and that is good.

Polarization can be used to cause certain phase relationships to exist in a light-ray-wave. If elements with optic axes are used to pass the beams, there are, essentially, a fast and slow axis of propagation in these crystals. This is governed by how the polarization of the light ray is oriented with respect to the optic axis of the crystal. A ray polarized parallel to the optic axis is said to be the slow ray, while the one that is polarized perpendicular to the optic axis is the fast ray (E ray). Thus, there can be specific phase relationships, or polarizations of the emerging rays, depending on time and distance of the wave passing through the crystals, and the wavelength-distance of the crystal element.

Retardation means to an electronics buff that the phase of the light ray has been caused to lag the fast axis propagation. Also, these plates can produce, with proper polarization changes, a circular polarization from a linearly polar-

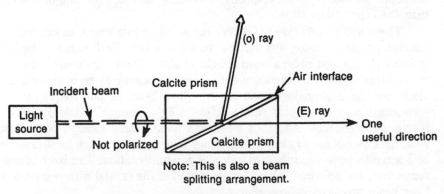

Note: This is also a beam
splitting arrangement.

Fig. 2-12. A polarizing prism passes the (E) ray.

ized beam. This is important in many applications. Recall that a laser produces linear polarization of its beam. To convert this to circular polarization, as required for ranging, for example, one uses retardation plates, crystals through which the beam must pass.

EXPLAINING POLARIZATION SYMBOLS

When drawing diagrams showing light beams and where they go and what happens to them, it is useful to use some symbology to indicate the polarization of rays or beams as they move through various elements of light control. Such symbology is shown in Fig. 2-13. The ray or beam of light is represented by an arrow with an arrowhead (vector) that shows the relative direction of propagation of the light ray or light beam. The polarization is indicated by drawing short lines or circles or combinations of short lines and circles (C), or just partial circles (rotating vectors) on the ray or beam lines.

By using this symbology it is easy to indicate how the polarization of rays or beams of light may change when going through the various blocks of a system of light control. In Fig. 2-13, the small cross-lines may be given arrowheads so they become small vectors to indicate whether the polarization is up or down as we see it graphically. This could be 90 degrees or 270 degrees, depending on the cross-arrow direction. Also, if the ray or beam vectors were drawn to scale, they would indicate by their length the relative

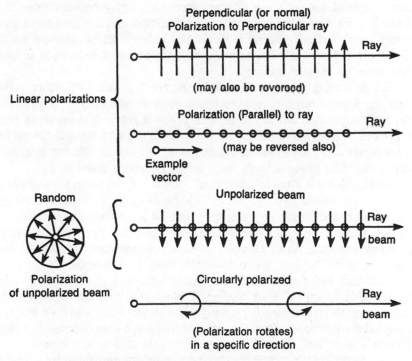

Fig. 2-13. Representations of lightbeam polarization symbology.

41

intensity or strength of the ray or beam. Sometimes the length aspect of the vector concept is not useful, so be careful.

USE OF PRISMS

There are many types of prisms and these can be used to control and direct the light of a laser beam.

The corner cube prism is designed by the manufacturer so that it returns a beam of light back toward its origin after three internal reflections, regardless of the orientation of the prism with respect to the beam of light. A 90-degree reflector prism is a right-angle wedge which deflects the light rays or beam 90 degrees from the incoming direction. The so-called roof prism is like the 90 degree prism in that it is a wedge shaped section of, say, silica glass. Actually it is an equilateral triangle, and if the light beam enters the base to one side it is reflected from the two sloping sides and comes back out the base some distance away from the entrance point. This type of prism reflects a ray or beam back towards its source, but unlike the corner cube prism the ray is spaced from the entrance line. It is not reflected back on itself, precisely.

A laser dispersing prism is one made from fused silica glass and polished. the laser beam is adjusted so it passes through at the Brewster angle and thus provides small loss with polarized light inputs. The beam goes into one side of the equilateral triangle, across the material, and out the other slanted side. The beam is spread out, as some different frequencies may be bent more than others. If a given laser has just one frequency component, it is bent just a given amount (refracted). If several monochromatic laser beams impinge on the prism (each different in color or frequency) the output is separation of those frequencies by physical spacing.

If a dispersing prism is cut in half, the result is called a Littrow prism. When the desired beam enters the hypotenuse, it reflects from the flat rear side, and comes back out in the same direction it entered. If the input beam happens to be from a multifrequency laser (there are such things!) the various rays separate and you can select the frequency of beam or light ray you desire quite easily. The prism is made from Schlieren quality fused glass.

Finally, there is a penta-prism that deflects a given beam 90 degrees no matter how it enters into the prism. This Penta-Prism is useful in the alignment of lasers and of auto-collimating systems. It is possible to rotate this type of prism and so create a plane of laser light that has some use in leveling operations and such. As an example, let's use the construction of a room for a large computer. The floor had to be exactly level. To accomplish this, a device was used that had a gyroscope that was self-orienting if left running a few minutes prior to use. When it leveled, it adjusted a laser that flashed a light beam around the room, clearly marking a line on the walls. The floor was then measured down precisely from this line and that was a level flooring! In a sense this was a laser-light plane throughout the room used as a reference.

The beam splitting cube consists of two prisms separated by a thin air interface, as previously mentioned, and the dove prism rotates an image about the beam axis by an angle equal to twice the angle of the prism as measured to the beam. If you are interested in these or other special types of prisms, Oriel, at 15 Market Street, Stanford, Conn. 06902 is a good source.

A DIAMOND IS A WINDOW?

A diamond provides an excellent window in a wide-band sense from about 0.23 μm to about 200 μm (micrometer) in the infrared band. The diamond is hard, durable, has high tensile strength, low thermal expansion, and high thermal conductivity, so it is excellent windows for high powered lasers. The diamonds used for this purpose are called type II diamonds. Of course other types of windows can be used, such as zinc selenide and germanium. The diamond or other compounds are ground into flat plates of very thin dimensions and then can be used to pass the beams of light out of chambers or tubes.

FILTERS FOR LASER LIGHT BEAMS

There are many types of filters that pass only certain amounts of light or pass just certain frequencies of light. These are important in light-control. The filter operates either by attenuating the light beam or by absorption of that light beam. It is possible to attenuate a light beam by passing it through an element that causes destructive phase relationships (polarizations) to exist which eliminate the output rays.

A reflective type of neutral filter can consist of a thin metallic coating on a clear substrate or base material. The amount of light passing through the coating depends on its thickness. Light that is not transmitted through the filter is partially absorbed and partially reflected. In any event if does not get through the filter. The purpose of the filter is to cause certain bands, frequencies, or colors of light to be eliminated or not pass on to other elements. It can also be used to reinforce certain frequencies or colors. The substrate material is usually glass or UV-grade fused silica. An absorption type substrate is usually a grey-colored glass and its thickness determines the amount of absorption.

Neutral density filters are used to fill a need for optical light attenuators as used in photomultipliers, radiometers, thermopiles, and such instruments. These kinds of instruments are usually sensitive to many frequencies of light and it is desired that the responses be uniform over broad bands. The types of filters used help to smooth out the peaks and valleys of various wavelengths without omitting any of the wavelengths of the incident light.

Sometimes various types of filters are used the same way in photography. (A study of camera photography and filters will help you to understand their types and uses in light and color (frequency) control of light rays and beams.) It is easy to mount such filters on a color wheel. Turning the wheel places various filters in the beam path and the effect of the filter can be studied. The content of the beam's spectrum can be easily analyzed in this manner.

NARROW BAND FILTERS

These can be likened to the filters used in radio receivers. They pass only a very narrow band of frequencies of the light beams and attenuate all others.

They are like the i-f system in a superhetrodyne in the way they work with light. The basis of this filter is a series of thin-film partially-reflecting layers arranged as a single or multiple Fabry-Perot interferometer. This interferometer causes destructive annihilation of undesired light rays within itself. You can read about Michelson's interferometer in a good college physics text.

In any event, considering the multiple layer interferometer, the spacings between layers is chosen so that beams produced by the multiple reflections from these layers are in phase with the transmitted wavelength for the desired frequency; thus, they go through the filter quite easily. Other wavelengths are not transmitted because of the destructive interference just described. By varying the reflectives, the number of layers, and the number of cavities in which the filters are placed (called the Fabry-Perot cavities), a wide variety of wavelengths and beam shapes can be produced. The center wavelength is always determined by the thickness of the thin film deposit.

FILTER TERMS AND DEFINITIONS

Light filters are described by various terms.

Peak wavelength. The wavelength of maximum transmission.

Central wavelength. Frequency halfway between half-power transmission values.

Peak transmission. A maximum transmission value.

Bandwidth. Width in wavelengths of the maximum value obtained between $\frac{1}{2}$ levels of maximum value. (Essentially, the $\frac{1}{2}$ db value as known in electronics.)

Band-Pass Shape Factor. A graph of the band-pass characteristics of the filter or filters.

Blocking Region. The wavelengths which are blocked by the filter.

Incidence Angle Shift Factor. Also called the effective index-of-refraction. Gives the change in peak wavelength with a change in filter angle of incidence. As incidence angle is increased, the peak wavelength becomes shorter. Changing the angle of incidence is also called tuning a filter.

There are many types and kinds of filters. This discussion is somewhat limited, but provides enough background so that other types may be readily understood, and their uses evaluated.

THE ETALON INTERFERENCE FILTER

You'll see this word as you read about the use of filters. It is an important device in laser light control and is a Fabry-Perot interferometer. It can be used to narrow the bandwidth of a transmitted beam (for example a dye laser, which generated a relatively broad band of wavelengths) by causing the undesired ones to interfere with each other destructively. The desired wavelengths then combine constructively within its reflective boundaries. When the Fabry-Perot interferometer is placed within a cavity that is tuned to some exact wavelength—so that a given beam frequency is produced—the interferometer is then called an Etalon Interference filter.

SOME BASIC TYPES OF LASERS

The uses of lasers are almost beyond imagination. In the medical field alone they are used for eye surgery, to "staple" retinas back into place when they become loose, to "weld" eye lenses back into place, and to cauterize wounds. Laser use also extends to such mundane subjects as photography, where laser color developers are using laser beams to create prints with near perfect fidelity using 35 mm color slides. In this application, three laser beams—one red, one green, and one blue—are used. In any event, these three beams are arranged so that they can scan the color slide or transparency and then, in turn, a receiver positioned on the opposite side receives signals that are proportional to the amount of each light color that passes through the transparency. These electronic signals (that come from opto-electronic sensors that are color sensitive) then go to a computer, which controls the emission of three identical lasers that are focused onto a photographic negative. This causes a print-type negative to be made from a transparency and color reproductions can easily be made from this print-negative, or C print. What is fascinating about this process is that the original art is scanned in such small detail that as many as 60,000,000 micromeasurements are actually made of the lighting of the original. This produces the remarkable detail! You'll immediately imagine that a person might alter the computer program somewhat and thus achieve special images and this is true. It can be done with one type of light emission circuit shown in Fig. 2-14.

THE SOLID-STATE LASER FOR EXPERIMENTS

Using proper precautions (you don't want to blind yourself)—you can do some experiments with a solid-state laser, assuming you know where to get one. Try Edmund Scientific Co., 101 Gloucester Pike, Barrington, NJ 08007. They also have optical fibers, lenses, and filters. To pump such a laser you might need a circuit using a transistor as shown in Fig. 2-15. This is a suggested circuit; monitor the voltage pulses across the laser resistor with an oscilloscope. Edmund Scientific may have other circuits which they prefer you use with their laser diodes. But be careful! Don't look at the light!

You need to experiment with the circuits to learn what currents and their durations are permitted for the laser diode you have obtained or plan to get. Once you know this, you can go about finding a suitable circuit to produce such current pulses. There are many circuits available. Care is necessary so you don't burn out your laser with too long a pulse or too high a voltage.

SOME LASER APPLICATION
INSTRUCTIONS FROM II-VI INCORPORATED

A peculiar name for a company, Two-Six Incorporated is located at Saxonburg Boulevard, Saxonburg, PA 16056. They are into lasers and manufacture infrared materials and optics. The infrared part of the spectrum is a good frequency range for laser operation. They have an interest in CO_2 lasers and provide polarization switches that can be used with high-powered pulse generators.

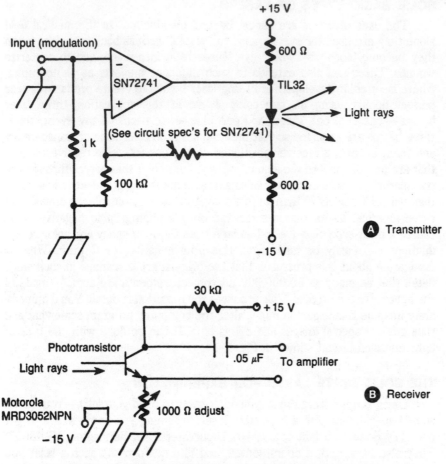

Fig. 2-14. Some output circuits to produce light from LEDs.

When using a fast, variable pulse length type of generator, it is now possible to electronically control both the optical turn-on and turn-off performance of a CO_2 type of laser. The switch is called a Q switch and it is used with a CdS quarterwave plate. Both are inserted into a CO_2 laser cavity that contains a polarization sensitive element such as a ZnSe Brewster plate. The laser will operate only when the polarization is favored by the Brewster plate. The quarterwave plate inserted in series with the Q switch contributes a round trip half-way retardation which effectively rotates the preferred polarization some 90 degrees and thus spoils the laser-resonator Q. This stops the laser operation quite quickly. But when the Q switch is pulsed to the quarter-wave voltage level, the unit then contributes a round-trip half-way retardation which nulls out the quarterwave plate component, and in turn, increases the Q of the laser cavity, turning it on. The laser remains on until the voltage is removed from the voltage-controlled switch. The repetition rate of the switch may be limited by the duty

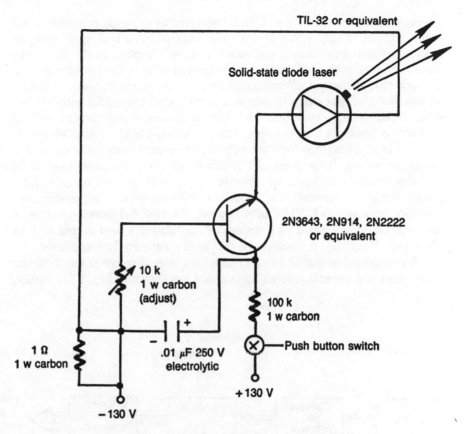

TIL-32 or equivalent

Solid-state diode laser

2N3643, 2N914, 2N2222
or equivalent

10 k
1 w carbon
(adjust)

100 k
1 w carbon

1 Ω
1 w carbon

.01 μF 250 V
electrolytic

Push button switch

+ 130 V

− 130 V

Fig. 2-15. A "pumping" circuit for a solid-state laser.

cycle restriction of the pulse generator or the power handling capability of the Q-switch electo-optic crystal. This crystal is a type CdTe, so its power capability can be increased by water cooling (desirable under some operating situations). A diagram of a switching system for a TEA laser is shown in Fig. 2-16 and details of the 3 EOSI switch in Fig. 2-17.

HOW DO YOU MODULATE A LASER?

The modulation system changes the phase characteristics of an incident laser beam and then changes the laser's performance based on that change. You pass the laser beam through a unit which is called a modulator and control something peculiar to the beam, its phase, frequency, polarization, amplitude, intensity or something in accord with some preplanned intelligent variation. Essentially this is modulation of anything.

In the case of amplitude modulation external to the laser cavity, the incoming vertically polarized light from the laser beam is passed in sequence through a series-arranged quarterwave plate. When no voltage is applied to

47

this plate or crystal the incident light is converted to a circular polarization and so the amount of light which passes on is reduced to 50 percent of what its maximum might be. When a modulating voltage is applied to the plate, this causes a change in the polarization to a horizontal value. If the output elements are such that horizontal polarization is best for transmission, you can readily surmise that the light intensity increases at the output toward full output value. Also, if the voltage applied is such that it causes a still further shift in polarization toward a direction that is not acceptable to the output elements, then the light diminishes from the circularly polarized (nominal) value and is less in the output. It turns out that practical limitations for linear modulation with this system may be on the order of the 20 to 80 percent range with 50 percent being the nominal value. Modulator units for lasers are available from manufacturers such as II-VI Co. of Saxonburg, PA 16056. Figure 2-18 shows a basic schematic of a modulating system for modulating a laser output with an FM signal. Figure 2-19 shows some details of frequency tuning a laser.

A general-purpose modulating system for a laser is shown in Fig. 2-20 and some suggested general purpose applications are shown in Fig. 2-21. Finally,

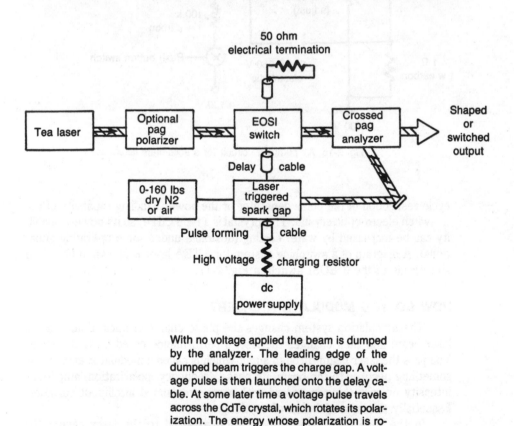

With no voltage applied the beam is dumped by the analyzer. The leading edge of the dumped beam triggers the charge gap. A voltage pulse is then launched onto the delay cable. At some later time a voltage pulse travels across the CdTe crystal, which rotates its polarization. The energy whose polarization is rotated then passes through the analyzer.

Fig. 2-16. A switching system for a TEA laser (courtesy of II-IV Co.).

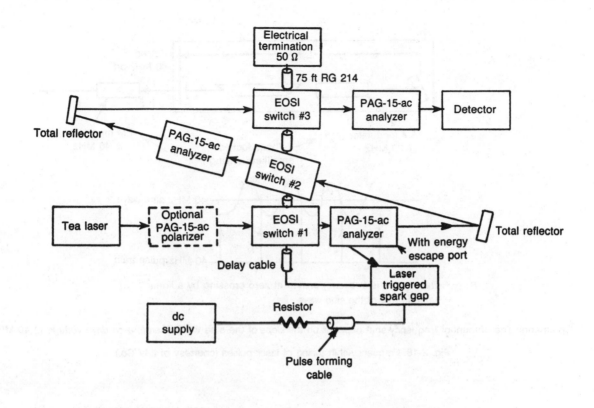

Case 1:	All crystals are pulsed to $V_{\lambda/2}$, and timing is adjusted to transmit the entire pulse through each crystal.
Result:	Cubed extinction ratio.
Case 2:	All crystals are pulsed to $V_{\lambda/2}$, the 2nd crystal timing is adjusted to cut off a portion of the pulse from the first crystal, the 3rd crystal improves the extinction ratio.
Result:	Pulses as short as 150 picoseconds require spark gap adjustment.
Case 3:	The 1st crystal is pulsed to $3\,V_{\lambda/2}$ then 3 short pulses are produced. The 2nd and 3rd pulse are pulsed to $V_{\lambda/2}$ and timing is adjusted to cut off all but the 1st pulse.
Result:	Pulses as short as 90 picoseconds require spark gap adjustment which here is critical.

Fig. 2-17. A laser-triggered pulse-switching system (courtesy of II-IV Co.).

Example: Given a mode locked laser with output pulse at 20 MHz rate. Want to frequency shift output pulses.

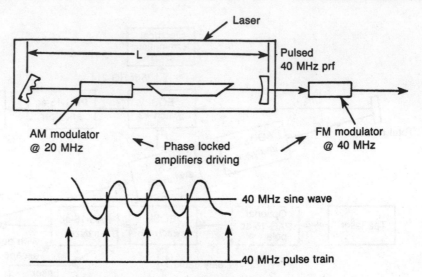

Each pulse is *frequency shifted* at zero crossing by a linear voltage rise of the sine wave

Comments: The amount of frequency shift depends on the *slope* of the sine wave, *therefore on drive voltage at 40 MHz*

Fig. 2-18. Frequency-shift tuning of laser pulses (courtesy of II-IV Co.).

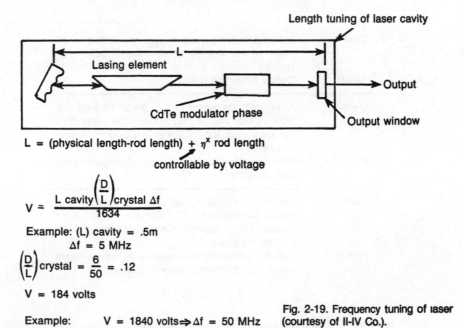

L = (physical length-rod length) + η^x rod length

controllable by voltage

$$V \approx \frac{L \text{ cavity} \left(\frac{D}{L}\right) \text{crystal } \Delta f}{1634}$$

Example: (L) cavity = .5m
Δf = 5 MHz

$\left(\frac{D}{L}\right)$ crystal = $\frac{6}{50}$ = .12

V = 184 volts

Example: V = 1840 volts ⇒ Δf = 50 MHz

Fig. 2-19. Frequency tuning of laser (courtesy of II-IV Co.).

50

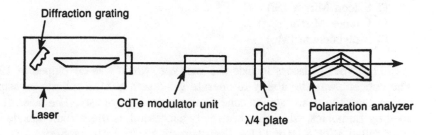

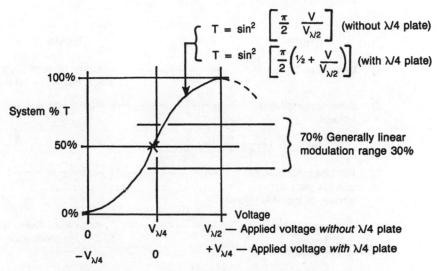

$T = \sin^2\left[\dfrac{\pi}{2}\ \dfrac{V}{V_{\lambda/2}}\right]$ (without λ/4 plate)

$T = \sin^2\left[\dfrac{\pi}{2}\left(\tfrac{1}{2} + \dfrac{V}{V_{\lambda/2}}\right)\right]$ (with λ/4 plate)

Fig. 2-20. General purpose modulation system (courtesy of II-IV Co.).

Fig. 2-22 shows some additional modulation concepts, and some details of an electro-optic switch in Fig. 2-23.

INTERPRETING SOME CHEMICAL SYMBOLS

In laser and optical fiber discussions, there are some abbreviations and chemical designations used that may not be as familiar as pnp and npn. For example:

☐ GaAs (gallium arsenide)
☐ Ge (germanium)
☐ CdTe (cadmium telluride)
☐ ZnSe (zinc selenide)
☐ ZnS (zinc sulfide)
☐ GaAlAs (gallium-aluminum-arsenide)

Since there are many types of lenses and mirrors used in this field the following may also be helpful.

☐ Silicon Mirror (Si)
☐ Copper Mirror (Cu)
☐ Molybdenum (Mo)

The types of lasers include the gas-type, which was developed in 1968. The concept was the ability to operate gas lasers at atmospheric pressures, rather than the near vacuum conditions required previously. The laser developed by Lumonics, which has been very successful, is the carbon dioxide CO_2 laser called a TEA laser (TEA Transversely Excited Atmospheric).

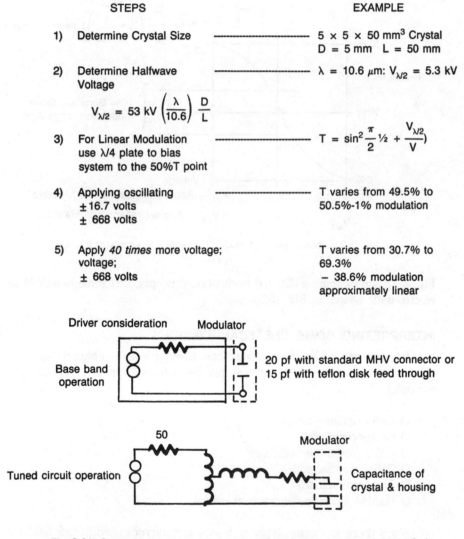

STEPS		EXAMPLE
1)	Determine Crystal Size	$5 \times 5 \times 50$ mm^3 Crystal D = 5 mm L = 50 mm
2)	Determine Halfwave Voltage $V_{\lambda/2} = 53 \text{ kV} \left(\dfrac{\lambda}{10.6}\right) \dfrac{D}{L}$	$\lambda = 10.6\ \mu m$: $V_{\lambda/2} = 5.3$ kV
3)	For Linear Modulation use $\lambda/4$ plate to bias system to the 50%T point	$T = \sin^2 \dfrac{\pi}{2}\frac{1}{2} + \dfrac{V_{\lambda/2}}{V})$
4)	Applying oscillating ± 16.7 volts ± 668 volts	T varies from 49.5% to 50.5%-1% modulation
5)	Apply *40 times* more voltage; voltage; ± 668 volts	T varies from 30.7% to 69.3% − 38.6% modulation approximately linear

Driver consideration   Modulator

Base band operation

20 pf with standard MHV connector or 15 pf with teflon disk feed through

50

Tuned circuit operation

Modulator

Capacitance of crystal & housing

Fig. 2-21. General purpose applications in AM procedure (courtesy of II-IV Co.).

A loss-modulation approach

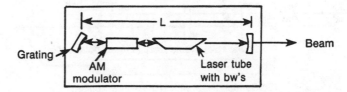

Comments
— place modulator unit at the grating end where the beam is smaller.
— Drive the modulator at a Freq = ½ cavity mode spacing

$$= \frac{1}{2} \times \frac{C}{2L}$$

— *Modulator causes cavity to exhibit high loss* except for the energy in modulator when the voltage crosses zero. After energy makes a round trip in cavity, it again sees *no loss* because of C/4L modulation rate, so the pulse builds up.
— The pulse width depends on the depth of modulation. The higher the modulation voltage the shorter are pulses produced.

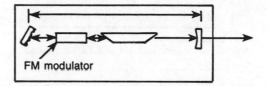

FM modulator

Comments
— Place modulator unit at the grating end where the beam is smaller
— Drive the modulator at a Frequency = $\frac{C}{2L}$

— The modulator shifts the frequency of the energy-out, except that energy which passes through the modulator as the voltage crosses zero. Again the pulse builds up because of the C/2L rate.
— The pulse (duration) depends on the driving voltage.

Fig. 2-22. Some added modulation concepts for lasers (courtesy of II-IV co.).

HOW IS A GAS LASER CONSTRUCTED AND HOW DOES IT OPERATE?

The gas laser circulates a certain type of gas through a sealed cavity chamber by using a pump. The gas passes over two oppositely charged electrodes which are energized by the high voltage power supply. The electrostatic field between these electrodes and the spark causes the atoms of the gas to become excited and rise to a higher orbit level. Since the gas moves in the chamber, the excitement of the atoms drops as they move away from the electrodes. When they fall back to their normal state, they give off light photons in a lasing beam. This beam oscillates in the chamber between mirrors until it reaches a highly sustained and usable level. Then it passes out through a suitable window in the chamber to be focused by proper lenses and used in

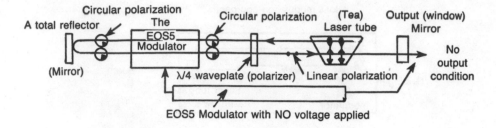

EOS5 Modulator with NO voltage applied

After passing twice through the $\lambda/4$ plate, polarization is rotated 90°; *Brewster windows* cause the cavity to be HIGH LOSS (LOW Q) for this polarization. No output results.

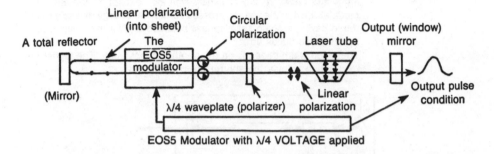

EOS5 Modulator with $\lambda/4$ VOLTAGE applied

Beam passing twice through the modulator with a $V_{\lambda/4}$ applied voltage rotates the polarization an additional 90°, BACK TO LOW LOSS (HIGH Q) state. So pulse output is present at the switching rate.

NOTES

The voltage rise-time must be faster than cavity-build-up time (typically < 50 nsec).
The repetition rate generally is limited by the duty cycle of the *pulse generator*.
A typical generator is very sensitive to load capacitance. That is the reason for the *flying leads* and *dielectric housing*.

Fig. 2-23. Details of the II-IV electro-optics switch (courtesy of II-IV Co.).

whatever manner is desired. Figure 2-24 illustrates in crude form such a laser. Notice the internal mirror system that makes the beam oscillate in the laser head and the window output where the beam is emitted. Of course, the gas (the working element) gets hot and normally it has to be cooled by a heat exchanger. There are some parts of this system not shown but you can imagine where they apply. A suitable working gas is one composed of carbon dioxide, nitrogen, and helium in a proper mixture.

It is important that the plane of polarization of a laser beam be properly oriented when the beam is used in a cutting application. When the plane of polarization is parallel to the direction of travel of the material the cut has the straightest sides. If the direction of movement is at some angle to the plane of polarization, a wide cut results with slanted sides. When the angle approaches

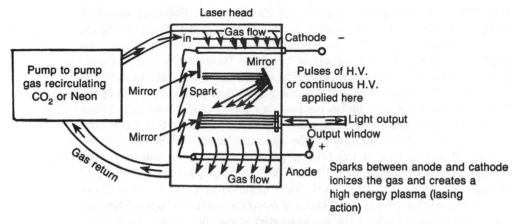

Laser head

in — Gas flow — Cathode −

Mirror

Pump to pump gas recirculating CO_2 or Neon

Mirror   Spark

Pulses of H.V. or continuous H.V. applied here

Mirror

Light output

Output window

+

Gas return

Gas flow   Anode

Sparks between anode and cathode ionizes the gas and creates a high energy plasma (lasing action)

Fig. 2-24. A gas-type laser.

90 degrees from the polarization plane, the cut is wide, ragged, and have rounded sides.

The ruby laser operates in the infrared part of the light spectrum. The gas lasers we are considering can operate in the infrared and also the ultraviolet portion of the light spectrum. All types of lasers can produce coherent light and are able to concentrate the energy contained in their beams in a small area to create high temperatures.

There are other laser types. There range from the rare gas halide (excimer) laser developed by Lumonics in 1978 to the minixenon chloride laser (a nitrogen type of laser), from the silicon marker laser (can mark silicon seminconductor wafers) to the dye laser that uses tunable light in the visible spectrum. In some cases, an ultraviolet flash source using its own electrical supply system may be used to pre-ionize a discharge cavity prior to the time that the main spark pulse is delivered to the cathode and anode in the lasing chamber. (Neon-helium and carbon dioxide mixtures have been very successful in gas lasers.)

WHAT IS A PYRO DETECTOR?

A pyro detector is designed to make pulsed energy measurements of laser systems of various kinds. It is capable of handling high energy pulses without damage. The detector element is constructed from a high Curie-point material and this is protected by ruggedized case. The detector delivers a voltage pulse whose amplitude is proportional to the number of joules impinging on its surface. The output is independent of beam width or its pulse length.

OPERATION OF THE LUMONICS TEA-820

As suggested in Figs. 2-24 and 2-25, such a laser is based on a pair of transverse-profiled electrodes providing a spark volume discharge of greater than 300 cubic centimeters, housed in a 25 centimeter fiberglass vessel. A rigid structure of rods hold the optical components inside the vessel. Lasing gas (a gas capable of population inversion excitement and relaxation and thus production of a laser light output) is circulated transversely over the electrodes. While this is happening, electrical energy from a charged capacitor is periodically discharged through the spark gap into the gas. That produces the population inversion and the laser action that produces the laser light beam.

The excimer laser is very much like the TEA laser except that it produces light in the ultraviolet portion of the spectrum instead of in the infrared portion. It is also good at this time to get some idea of the pulsing energy needed to operate such a laser. In the TEA 600A, CO_2 gas laser, a charging voltage of up to 50 kV is needed. The operation of some of these lasers requires as much as 800 to 1000 joules of electrical energy to be delivered into the discharge cavity where the gas is located, during one microsecond pulse. This can require from 140 to 160 kV of electrical energy and a maximum of

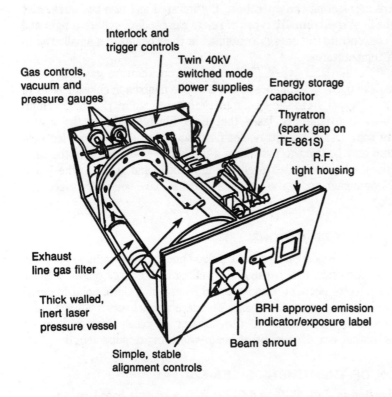

High average power:
18 watts, KrF, 249 nm
10 watts, ArF, 193 nm
8 watts, XeCl, 308 nm
1 watt, F_2, 157 nm
High pulse rates:
To 200|pps
High pulse energy:
To 250 mJ (KrF, 249 nm)
Extended gas lifetime:
Guaranteed 2×10^6 shots to 50% power (XeCl, 150 Hz, 308 nm)
Multigas operation:
Includes all normal RGH transitions, plus N_2+, N_2, F, F_2, CO_2, etc.
Single cabinet construction:
Convenient single table-top housing including all gas and electrical controls, power supply, gas filter, etc.
Low RFI:
Single shielded cabinet and careful design results in less than 5 mV rms RFI (TE-861T, TE-861M)
Low jitter:
±2ns jitter (TE-861T, TE-861M) relative to external input trigger pulse
High wall-plug efficiency:
Only single phase, 120 V line required (optional 240 V)

Interlock and trigger controls

Twin 40kV switched mode power supplies

Gas controls, vacuum and pressure gauges

Energy storage capacitor

Thyratron (spark gap on TE-861S)

R.F. tight housing

Exhaust line gas filter

Thick walled, inert laser pressure vessel

Simple, stable alignment controls

Beam shroud

BRH approved emission indicator/exposure label

Fig. 2-25. A typical lumonics gas laser (courtesy of Lumonics).

56

inductance in the energy storage network. In a rare-gas laser, halogen gas and helium are mixed and supplied to the laser cavity. Another rare type of gas is xenon. Laser cavities are usually pumped with a vacuum pump to clear out discharged gases.

Finally, if you'd like a laser kit to experiment with, you can write to: Lumonics at 105 Schneider Rd., Kanata (Ottawa) Ontario, K2k 1Y3. Figure 2-26 shows the gas-handling details of the TE 262 laser.

Figure 2-27 shows the schematic of a dye laser. Notice that an excimer pump (ultraviolet output) laser is used to excite this dye laser. Also, note that a diffraction grating is used in the tuning mechanism. The excimer pumping beam goes to the dye cell oscillator (3) where the laser beam is formed. The beam then goes into the tuning mechanism for reinforcement by diffraction grating adjustment at near grazing incidence. A low magnification (15X), four-prism achromatic beam expander is used to get high conversion efficiency and narrow and constant line width on the output beam which then passes to the output coupler (4). What happens here is a positive type of feedback at the desired frequency through the dye cell reinforcing the energy until the beam is of sufficient intensity to be used. It is then coupled through an amplifier dye cell. The amplifier dye cell also is pumped (master action) by the excimer beam.

OPERATION AND DETAILS OF LASER WAVEPLATES

Since laser outputs are light rays, we are concerned with focusing of these rays, the polarization of the rays, and the frequency or wavelength, and reflection of these rays. In fact, we are concerned with all elements which enable us to control, manipulate, direct, and use the light rays to our advantage. One such element is the waveplate.

When a plane polarized beam is incident on a birefringent plate whose optic axis lies in the plane of the plate, the beam is resolved into two components. These components propagate through the plate with different velocities

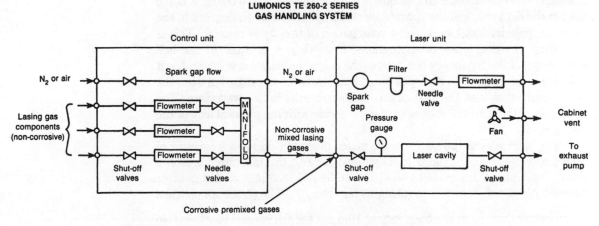

Fig. 2-26. Engineering diagram of a gas-flow laser (courtesy of Lumonics).

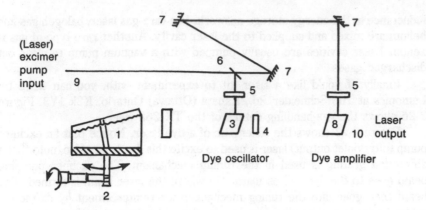

1. Tuning mechanism
   (with four-prism expander)
2. Sine drive
3. Oscillator dye cell
4. Output coupler
5. Cylindrical lenses
6. Pump beam splitter
7. Pump beam steering mirrors and
   optical delay line
8. Amplifier dye cell
9. Excimer pump input
10. Dye beam output

Fig. 2-27. The schematic operation of a dye laser (courtesy of Lumonics).

and recombine upon exiting. There they have a new and different polarization. The character of this polarization depends on the orientation of the optic axis and also on the net difference in optical path length of each of the two components. This governs the phase relationships of the two components of the wave and thus the resultant phase of the combined wave. This difference in optical path length (d) is given by an equation:

$$(d) = \Delta nt$$

where Δn is the birefringence of the material, and t is the thickness of the material. When d is equal to one quarter wavelength of the frequency being used in the light ray, and the optical axis of the material is at 45 degrees to the incoming polarization, the outgoing polarization of the ray is circular. This is necessary in some types of applications. But, when d is equal to one-half wavelength of the frequency being used, the output polarization is linear, but it is rotated with respect to the input polarization by twice the angle between the input polarization and the optical axis of the material in the waveplate. The optic axis of the material may or may not coincide with the physical axis of the material.

The thickness of the quarterwave plates made of CdS for a frequency of 10.6 μm is very thin, making it fragile and hard to manufacture (.1 inches thick). Such plates are made by II-VI Inc. to be odd multiples of that quarter wavelength (3/4, 5/4, or 7/4 wavelengths for strength.) The plates are polished flat to better than $\frac{\lambda}{40}$ at the frequency of 10.6 μm and are parallel to within two seconds of arc. They have an antireflection coating applied at the designed wavelength. Some of these waveplates have a transmissibility factor of as high as 97 percent.

If you retard a light ray (as is often mentioned in the current literature) you actually change the phase or polarization of that light ray. Often it is necessary to have some variable phase controls over the light ray and for this purpose an element called the Babinet-Soleil Compensator is used. The compensator is constructed of two wedge-shaped plates which are oppositely oriented. This produces a variable multiple order plate. When another fixed thickness parallel plate with its optic axis is rotated 90 degrees is added to the two, then you have a variable equivalent order zero plate.

The compensator consists of two CdS wedges with the wedge angles in opposite directions. One such wedge is adjusted by a micrometer in a transitory manner in the direction of the wedge. Thus the two wedges have a variable thickness whose birefringence is offset by an added parallel plate. The net effect is to produce a variable retardation, adjustable by the micrometer setting. Such a device is adaptable to beam splitting types of operations.

BEAM SPLITTING AND LASER-RANGING MECHANICS

The separation of various polarizations of the save wavelengths of a laser beam is often accomplished by using a Brewster window (see Fig. 2-28). Essentially, one half of the incident beam is bounced off the reflective surface with a polarization change. The second half of the incident beam goes through the material with essentially the same polarization as the incident beam. Thus the incident beam is split into two components whose phase is somewhat

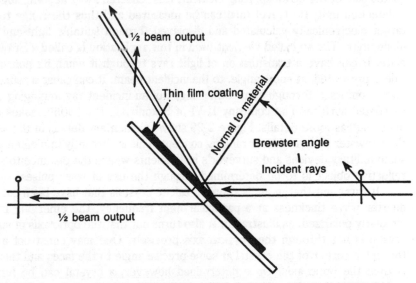

Fig. 2-28. A TFP (thin film polarizer) beam splitting concept (courtesy of II-IV Co.).

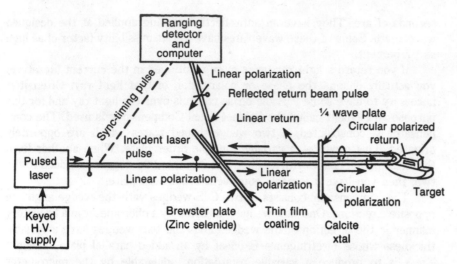

Fig. 2-29. A laser ranging system concept (courtesy of II-IV Co.).

different. If the rays are later recombined, the cancellation or reinforcement effect can be measured and used constructively.

In the field of laser ranging, such a device can be used in the following manner: A "P" polarized pulse is sent out through the beam splitter, through a phase adjuster (retarder) to convert the linear polarization to a circular polarization and then the laser ray goes on to impact the target. Assuming reflectivity, the ray is bounced off the target and returns, again through the wave (phase) adjuster so that it is converted into a linearly polarized wave whose phase is slightly different from the incident ray. Upon impact with the thin film coated side of the beam splitting element, it is reflected away at some angle to a detection unit. Its travel time can be measured and thus the range to the target electronically calculated and displayed through suitable light-emitting diode units. The so called element used in this application is called a Brewster plate. It can have a transmission of light rays through it when its noncoated side is presented, at some angle, to the incident beam. It can cause a reflection away from itself through some other angle, of an incident ray impinging upon its coated surface. The company II-VI of Saxonburg, PA 16056 makes such units and has more details. Figure 2-29 shows elementary details of the use of the Brewster plate in a laser ranging concept. This is currently implemented in some military devices and surveyor's instruments where the distance to some reflective object is to be determined through the use of laser pulses.

It turns out that in applications of calcite crystals that have been cut to a quarter wave thickness at a particular light frequency, the emerging ray is circularly polarized, as illustrated. It also turns out that the optic axis of such a crystal is not through the physical axis precisely. One may construct a line through a corner of the crystal at some precise angle to the faces and this line is then the optic axis. Once determined however, a crystal can be further machined so that the optic axis will coincide with the physical axis if desired.

60

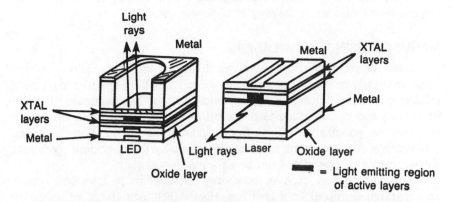

Chapter 3

= Light emitting region
of active layers

Applications of Lasers and Laser Communications

APPLICATIONS OF LASERS

There are possible military applications of lasers. Some of the laser types considered for this use are the hot-gas lasers in which a gas is heated to a high temperature, exciting the molecules, and then abruptly cooled. This amounts to a pumping situation of the atoms to achieve a laser output. Passing a gas through an electric field can cause the excitement among atoms which is necessary to make them radiate light photons in large numbers. Once the lasing action begins, it is a relatively simple manner to cause the small initial energy beam to regenerate back and forth in the gas or confined area (by means of mirrors) until the energy gets large enough to pass-out of the cavity or enclosure through a partially reflecting window. In a military application, the desired use is almost always one of destruction. Ranging, aiming control, and position control certainly are usable concepts in the application of laser light beams.

Damage resulting from a laser beam hitting a given target is due to the enormous amount of heat generated at the point of impact. Damage is caused only by that fraction of energy absorbed by the target. If the target reflects the beam and does this efficiently, the amount of absorption is minimized and damage capability is vastly reduced. The absorption of 1,000 watts/cm^2 or about 1,000 joules per cm^2 can melt a metal surface a few millimeters (hundredths of a meter) thick. A pulsed beam of laser light energy might cause evaporation of the metallic surface of a target. The action equals reaction concept of the metal flying away from the target might produce a reverse strain and tear a target apart. But to produce such energy on a target would require lasers capable of megawatts of power output.

MANUFACTURING WITH LASERS

Laser beams have some properties that other light sources do not have. They are nearly or exactly monochromatic, and they have spatial and temporal coherence. (That simply means that such a beam has continuing in-phase properties and polarizations as it travels away from its generating source, assuming no polarization effects from windows and such.) It is this spatial coherence of the laser beam which makes it suitable for machining operations. These machining operations can be on a micro scale.

In a modern laser, most of the energy contained in the laser beam can be concentrated (focused) to a spot less than 0.001 inch (0.03 millimeter) in diameter when using ordinary focusing lenses. What governs what happens when we focus a laser beam on some kind of material? First, consider the power density in the spot and second, how long the spot is applied to the material. Also, the frequency of the laser light and the kind of material upon which the beam is focused must be considered. When you concentrate the energy on some material (just like the magnifying glass and the sunlight) the point of concentration heats to a higher temperature. This may result in ablation (recall the ablation of the heat shield of the space shuttle) which is the total removal of the material within the focus area. The amount of energy concentrated gives rise to an ability to trim some material, scribe on some material, cut some material, or weld some material. In any event, the operation must be held within the focus range of the lens system to get the most concentration of the energy present. In micromachining, lasers are used that have wavelengths of from 10 to 0.5 microns, visualize this as going down scale on a graph from the longer wavelengths to the shorter wavelengths. It is said that wavelengths from infrared to ultraviolet are usable in manufacturing processes.

LASER EQUIPMENT FOR MANUFACTURING PURPOSES

Of course, you need a suitable laser, power supply, and cooling devices. You need to be able to switch the beam off and on at precise moments. You need to be able to position the beam precisely. You probably need a feedback system to monitor the operation and inform some controller mechanism when the operation is completed. You also need some safety system to prevent damage.

Because certain materials require certain wavelengths of light and certain powers of concentration for various operations, many different lasers, might be required in a factory. Lasers have found application in industrial applications in the welding and soldering of electronic parts. They are also used for heat treatment of various machine parts and for drilling holes, which in some applications may be from 0.05 inch to as small as 0.007 inch in diameter. Since the laser beam is narrow, it can be focused even more so that its point of concentrated energy is very small.

Lasers are in the high-power or low-power class. This is like saying that the machinery is "light" for low powers and "heavy" for the high powered devices. The dividing line may be on the order of ten watts output. Many solid-state lasers that produce the smaller outputs. They do produce sufficient energy to cut through materials such as metal, cloth, and plastics. They even shape ceramics used in the electronics trades.

Ruby lasers are also used in manufacturing. One end of the ruby rod is a one-way mirror and the other end is a pass-through window-mirror. When the light flash stimulates the atoms in the ruby rod, the light energy flashes back and forth inside the rod until it gains sufficient energy to pass out the window end of the rod through the focusing elements. Ruby lasers used in some manufacturing applications operate on about 0.69 micrometer wavelength. There are other solid-state types like the neodyminium-doped yttrium-aluminum-garnet lasers that operate on about 1.06 micrometer wavelength. These wavelengths are in the infrared region.

A ruby or garnet laser can operate with pulsed light flashes to heat a work surface. Other types such as neodymium-doped yttrium-aluminum garnet types may also deliver a steady or pulsed beam of laser light as desired. This latter type can be pulsed at a fast enough rate so that the modulation of the laser beam can be used for communication or data transmission applications.

Gas lasers such as argon and carbon dioxide types operate well in a continuous fashion because they have cooling systems. Some gas systems such as those envisioned for military applications such as the heated-gas (flourine) and electronically charged gas-discharge types are not considered for manufacturing applications in general. A survey has found that many "heavy duty" and high-power lasers are of the carbon dioxide type and can be operated continuously.

The laser beam delivers a highly concentrated amount of energy in a short space of time to a very small area. In many situations this is important, as it decreases absorption of heat by the remaining material of the workpiece. Thus less damage occurs on the balance of the work area of the material. This may eliminate cracks, metallic distortion, or the setting up of heat stresses in materials which sometimes happens when the material is subjected to broad applications of heat to its surfaces. Also, due to the high concentration of energy, the material to be removed from a cut or a hole may be vaporized leaving no extraneous pieces of material to be removed.

In some applications a laser beam is connected with an oxygen supply. When the heat is generated by the beam the oxygen is applied. The result is a burning of the material which produces fast cutting or shaping of the material. Suppose you want to deposit a material to some surface. Using a laser beam and feeding the alloy at the proper rate, the beam can heat it to the melting point and deposit it upon the receiving surface. Since the laser energy is well concentrated, very little heat is delivered to any other area. This operation is a point source type of application.

There are innumerable applications of laser cauterization in medicine. Tacking tissues together as in eye operations are now quite common, and the danger is reduced because of the fine plane-of-focus necessary to concentrate the light energy. Daily we learn of new applications in almost all fields of the use of laser beams.

In the field of medicine, spectacular developments seem to be under way using micro-size lasers. Some researchers have used tiny lasers attached to catheters to remove fat obstructions in the arteries of animals and do some repairs of the values in the heart. The marriage of camera-like fiberoptics to small optical fibers inserted into catheters enable medical researchers to probe deep inside the human body and examine many areas heretofore visible only after extensive surgery.

Lasers are used to open tissue and seal it. The micro-miniature laser types and the catheters now available presage a time when internal surgery might be accomplished without opening the body. Reduction of infection and smaller incisions are the spinoffs of using laser technology in medicine.

One patented method of tuning a laser uses a piezoelectric crystal. The patent describes the crystal as having a reflective surface capable of being deformed with electro-acoustic waves. The deformation of the surfaces causes the crystal to act like a diffraction grating and thus filter beam wavelengths so that various ones might be amplified.

Switches for laser beams have been made from acoustic-optical cells and their associated optical components. A fast chirp (changing frequency wave) in the cell causes the light beam to be diffracted and therefore matches by the transfer function of the optical system. When the diffraction is not precise the optical match does not take place and so a Q switch or very fast optical-shutter effect takes place. A high-power Q switch has been developed. This switch has been used in a 100 megawatt ruby laser cavity. A Frustrated Total Internal Reflector (FTIR) is pulse driven by a mechanical shock-wave to produce the fast switching action.

The steering or positioning of a laser beam is desirable in many applications. You want to change its direction as a function of some other quantity so that it will perform some particular task. The question, then, is "How do you steer a laser beam?" The Hughes Research Labs of California has come up with a method. They have developed an electronic beam scanner for a CO_2 laser to meet the following scientific requirements. They wanted a 3 degree scanning range: 100 resolvable beam positions (think how small a measurement that would have to be for a maximum scan of 3 degrees); a 6,000 Hz raster scan in one dimension; and a capability for 100 watts of 10.6 micrometer wavelength output.

In the device developed, they employed an optical Bragg diffraction grating germanium, with an acoustic-longitudinal wave, and a laser beam polarization in the crystal (111) direction. The numbers refer to the crystal's axes. An FM sawtooth wave centered near 100 MHz with a 27 MHz deviation was used to provide the scan signal and mechanism. A small transducer provided sufficient beam divergence in an acoustic mode to produce the 3 degree scan capability. Direct contact water cooling of the Ge crystal was provided to prevent thermal runaway. Cylindrical, reflective, telescopic optics were used for beam conditioning and shaping recollimation. The transducer bonding was done ultrasonically using thin metal film of Ag, Au, and Ag-In. The latter gave the best bonding performance.

In a program of study done by Hughes Research Labs at Malibu, California, an attempt was made to determine the best possible modulation system for CO_2 laser. This laser operated on 10.6 micrometers and was to be intensity modulated. That brings to mind immediately some kind of shutter or opaque-transparent arrangement or device that can be varied in ability to pass the laser beam as a function of some modulation current or voltage.

The Hughes performance requirements were a minimum frequency response of 200 MHz and a modulation capability of 100 percent with a minimum of driver power. Thus they began a serious review of all known modulation techniques for laser beams. This listing is a valuable resource when learning about laser modulation techniques. Some of the methods considered were the electro-optical effect, Stark Effect, acoustic-optical effect, and free-carrier related effects. The electro-optical and the acoustic-optical effects imply that the passage of the laser beam would be affected by either an electric current, which would affect an optical device, or an acoustic wave, which in turn was probably produced by an electric current varying within some controlled rate or manner. Of course, there were other techniques that they explored that either weren't successful or were of such a nature that divulgence of the process was not in their best interests.

Through an elimination process the techniques that did not have the 200 MHz modulation capability bandwidth and those that required too much power to make them effective were eliminated. This left some techniques that all embodied some kind of optical effect as controlled by electronics. These turned out to be an optical waveguide, a microwave bandpass traveling wave, and intercavity coupling methods that use microwave techniques. The changes may be induced electrostatic or electromagnetic fields, or physical changes in capacitance or inductance of the cavity, or merely its resonant size. Also, a baseband traveling-wave (TEM) parallel strip and multiple pass lumped elements were used. It was found that the intra-cavity coupling was very attractive and a flat bandpass in excess of 500 MHz was produced.

CONSIDERATION OF BRAGG'S LAW

Bragg's law predicts the conditions under which diffracted X-rays from a crystal are possible. As it turns out, a crystal (such as calcite) that can refract X-rays can be considered to have any number of refracting planes within its structure. The X-rays are considered to penetrate the crystal and to be refracted from these planes. The trick is to refract them so they add constructively and form a beam. Bragg was able to derive an equation that proved that the beam formation could occur if the spacing between the planes in the crystal were an integral number of wavelengths; i.e., (1), (2), (3) etc., and his equation (which reduces to a simple form) is:

$$2d \sin\theta = m\lambda$$

where d is related to the cell dimension by:

$$d = \frac{a_0}{\sqrt{5}}$$

When X-rays fall on a set of atomic planes so that the above equations are satisfied, then

$$\lambda = \frac{2d \sin\theta}{m}$$

where m is an integer and is the X-ray wavelength. Diffracted beams of X-rays result. Sodium chloride, for example, produces a refracted beam at θ equal to 12.6 degrees for m = 1.

OPTICAL SYSTEMS DATA TRANSMISSIONS

Assuming that you can modulate a laser beam satisfactorily with the required flat bandpass and depth of modulation required to convey signal data intelligence, what can you call this process? When two signals are combined, you have hetrodyning. Experimentation has been done to combine in multiple hetrodyning fashion microwave signals with infrared or optical signals. Such a data modulation system uses the multiple passes of the infrared or optical signals through a crystalline material which also has an acoustic-optical signal applied to it making the crystalline material into an acoustic-optic grating. Meantime, in the growing of the crystal and preparation thereof, fixed gratings are fashioned so that the optical signal or infrared signal is properly guided through the material. The output of the device is composed of the input signal (infrared or optical) that has become intensity modulated with the desired data to be transmitted. This signal can be transmitted optically, and here, optical fibers or optical-fiber cables may play an important role. The optical signal can be converted into an rf signal for transmission.

LIGHT BEAM SCANNERS

A laser beam that impacts with a piezoelectric crystal can be made to perform as a traveling-phase grating. This diffracts the monochromatic light beam. This traveling-phase effect is made possible by an acoustic-wave that propagates across the crystal to produce the grating phenomena. The light beam from the laser is diffracted in accordance with the acoustic wave deformation of the crystal. The light intensity at some first order spot is detected to provide a temporal signal representation of the light distribution of an image line. If the image is in colors, then by using a chromatic beam, one may separate the three primary wavelengths to re-create the image in the scanning beam.

It is time to reconsider this important light-control element. It is used so very much with laser beams, and we find it in so many diverse forms that a review of the grating is beneficial. When you have two slits, as in Young's experiment, assume that such slits are small compared to the wavelength of light passing through them, and so the light is diffracted rather uniformly (bent away from its original direction of propagation). Examine Fig. 3-1. The screen upon which the slit-beams fall has light and dark spots upon it. When the radiations in the beams are in phase, a reinforcement takes place and you have brightness. An out-of-phase condition darkens the screen.

A wider system of diffraction (bending) of the rays of light occurs if you use a spot source, lens, and a wider slit. If no lenses are used this is called the Fresnel diffraction. If lenses are used before and after the slit it is called a Fraunhofer diffraction. A larger number of "slits" (or as they are more commonly called "radiators") makes a diffraction grating. In this case, the larger number of slits enables the propagated rays to mix in phase or out of phase," or at some phase relationship between these extremes. The result is light and dark lines or spots on a screen (if one is used). The diffraction grating is a device that performs the equivalent operation of a multi-slit window for light rays.

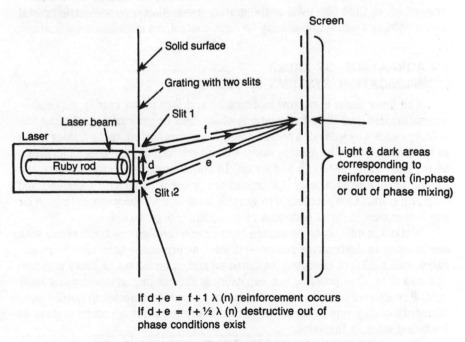

If d + e = f + 1 λ (n) reinforcement occurs
If d + e = f + ½ λ (n) destructive out of phase conditions exist

Fig. 3-1. A two-slit deffractor as per Young's experiment.

Thus you can imagine how using some kind of elements that provides this slit effect, controls the amount of light passing through it. Diffraction gratings can be made from glass that has been lined or scribed with fine lines of a proper depth and thickness so they act like slits. They can also be made of films or various other substances that can reflect or transmit light rays through them.

Not far from where I live there is a traffic signal that has a diffraction grating of a rather large size covering each light. As I approach the signal we cannot determine which light is on and which one is off. But, when I get close enough and in position near the intersection of the two roads, I suddenly find I can see red or green quite clearly. This is a diffraction grating system in front of the lights. It passes the color accurately, and passes them only to that distance wherein the grating-plus-lens is focused.

In any event, a diffraction grating is very important in light control. Gratings now can be varied so they pass more or less light as a function of some electric current or acoustic wave or such other phenomena as heat, pressure and so on. This makes it possible to develop signal systems, alarm and sensory systems, and convey data by amplitude modulation of the light beam generated from some laser source or LED.

Gratings are made by ruling equally spaced parallel lines and have depth. If a metal plate system is used it is called a reflection grating as light does not pass through it but is reflected with all the interference effects of the pass-through gratings. To cut the grooves precisely it is usually necessary to use a machine with a diamond cutting element. Once a grating has been made from some metal, then plastic films of various types may be coated on the face, stripped off and the film used as the grating itself. Also, a piezoelectric crystal can be used as a diffraction grating for light control in a modulation application.

AN ADVANTAGE OF LASER
COMMUNICATION SYSTEMS IS SECURITY

The laser beam is narrow in diameter and thus is not readily exposed to non-authorized and illegal revelations of its data or information. Among the military, such a system as a secure communications system using a laser beam as the carrier would be ideal in many areas: tank-to-tank, sub-to-sub, airplane-to-airplane, airplane-to-sub and so on. In some extremely busy harbors, the harbor master uses a laser communication system to control incoming and outgoing traffic. One problem is to get that laser beam to propagate through air and water over desired distances at sufficient energy levels.

While it is difficult to propagate laser beams through the atmosphere with any amount of destructive power—at least at present—they can be propagated with sufficient energy to be detected and can transmit or carry information and data. One proof of this capability is the ranging systems using laser light. For shorter communications systems internal to vehicles in hostile environments of any type, the use of optical fibers promises a secure system as discussed later in the book.

There is one phenomena of the laser beam which also helps to prevent security leaks in communications and data systems: the polarization effect. For various reasons, the polarization of a monochromatic beam of light may be fixed at some definite, relative position. It may then have to be rotated through some required angle in order to be receivable in proper form. Thus it may be possible that interception of a laser beam that may have some peculiar way of handling the polarization of its constituents as a function of time, message, or type of data may be somewhat difficult.

In this same sense, another concept relating to beam security is the tunable dye-laser that has an acoustic-optic beam deflector, a folded Lyot filter, and a diffraction grating that altogether produce a discrete, periodic, or digital output. The periods between the discrete wavelengths in the output of the laser are varied electronically with the input data or intelligence to be communicated. Dye-lasers can be operated continuously and thus provide a good basic means of continuous-wave communications.

In the quest for fast laser tuning, which means adjusting them for close-but-different wavelengths of output, an organic dye laser can be used. The tuning element may be a diffraction grating and a Bragg diffraction cell driven by an rf source, which modulates an ultrasonic transducer. Notice here the use of two elements previously discussed as important in light-control systems.

LASERS IN COMMUNICATIONS SYSTEM

The use of light beams for communications is a development that finds great favor among many scientists and researchers. The fact that the equivalent of a hard wire direct coupled system might be made that has no wires immediately brings to mind security possibilities such as in airplanes, battleships, submarines, on battlefields, or over distances that are so great that other types of communications cannot be easily installed, handled, or maintained.

A laser beam is a radiation in the electromagnetic spectrum. The use of modern modulation techniques conveys intelligence over such a link, and adaptations of normal communications equipment can be made to facilitate this communication. A case in point is the development of a 10.6 micron infrared superhetrodyne receiver front end for use in a wide-band CO_2 laser communications link. The infrared receiver used an 850 MHz response PV (HgCdTe) photomixer mounted in a suitable housing. An i-f amplifier and preamplifier was also used that had a response from 5 to 1500 MHz. The receiver was designed to accommodate a Doppler shift of some plus or minus 750 MHz while providing an instantaneous information bandwidth of some 400 MHz. This unit was tested by the Airborne Instrumentation Labs at Mellville NY. This demonstrates the use of the somewhat conventional types of ideas in fabricating an infrared communications receiver front end.

THE LASER IN A FIBEROPTIC STRAND

Optical-fiber lasers use distributed feedback to make them operate, and their use is envisioned for communications systems, which employ optics as the basis for the operation. At the National Aeronautics and Space Administration of Pasedena, California a laser was made of an integral part of the optical fiber channel either by diffusing active materials into the optical fiber itself or surrounding the optical fiber with active material. The oscillation of photons within the active medium to produce the lasing action (the pumping) was done by making a grating of the optical fiber in such a way that distributed feedback (positive reinforcing feedback) took place.

This is evidence of the joining of the fields of optical-fiber technology and laser technology to produce a device that embodies something of both technologies. This book has mentioned using a laser of small dimensions to feed an optical fiber, but here we find that the optical fiber itself may be adapted to being a laser and from this start we can easily imagine joining of more optical fibers to extend the range and use of the light thus generated. We easily imagine the possibility of modulating the laser beam with any pertinent kind of information and the reception and decoding of such information at the user end into usable form.

SOME PROBLEMS WITH LONG DISTANCE LASER COMMUNICATIONS

If you are trying to establish a communications system using a laser beam over a vast distance such as might exist between a satellite and a ground station, one of the first problems to be encountered is that of acquiring the target and keeping it in view so that the very narrow beam won't drift away from it. This may require rather complex instrumentation at each end of the system (for example, a radar to acquire the remote station initially). This radar might be a microwave type. Then a laser type of radar could be used to refine the pointing of the ground communications transmitting-generating system. You don't want to just illuminate a satellite with a communications laser beam, you want to illuminate directly and accurately a receiving sensor on that satellite. If you consider a laser beam to be $\frac{1}{4}$ inch in diameter and the satellite to be some 9 to 12 meters in size facing the radar, realize that a precise focus of the laser beam to one spot on that satellite's surface is required.

Some scanning probably has to be accomplished to acquire the target sensor or receiving sensor. Some kind of lock-on has to be employed through the various equipment at both ends of the laser beam channel, and then the beam can be modulated with the desired intelligence.

Some years ago a series of optical communications experiments between a target at 60,000 feet and a ground based station were conducted by NASA. The basic system was an optical tracker and a transmitter that were located at each end of the communications system's link (one airborne and one ground based.) The aircraft transmitter consisted of a 5 mW (NeHe) laser with a 30 megabit modulator. The ground station was an argon laser operating on a 488 nm wavelength. The problems of atmospheric degeneration of the system, acquisition, tracking, and such, were carefully studied and many experiments on data transmission were conducted.

Both speech and music were transmitted in excellent fashion over a few kilometers. But this experiment implies a successful modulation system, which with music of high fidelity must have a good bandpass, perhaps as high as 15 kHz. Speech sibilants require larger bandpasses if they are to be accurately and completely transmitted. Also the recovery system, the receiver, must be capable of demodulating such a modulation system with just as much precision. For this level of transmission (speech and music to video and pulse-data transmissions) much work is still required for the level of perfection desired.

In this connection, a system known as Pulse Quaternary Modulation (PQM) has been successfully demonstrated in laboratory experiments. The significant part of this modulation system was that an operational 400 Mbps bandbase laser system might be developed on this concept (Mbps is Megabits per second).

In a signal patent for laser communications, a device is described that can split the output beam from a laser into two quadrature polarized beams and then phase modulate one of these beams and recombine the beams in a noninterfering manner for transmission along a single path to the modulator. At that point the beams are received by a receiver and separated and then combining the beams destructively causes amplitude modulation of the combined beam. At the target, a receiver can remove the modulation information from the laser beam and send it to any kind of desired read-out unit. This is another example of how you can modulate the amplitude of a laser beam.

Back to the problem of laser communications over any long distance. Due to the narrow beam divergence of a laser or other optical communications system, very small angular motions of either end of the system can cause misalignments which could mean loss of data or intelligence. One location or situation in which this factor has to be considered in the military applications of laser beams for communications channels. In a moving tank there is a lot of bouncing around. In other vehicles, such as helicopters and airplanes, the wind forces on the body tend to cause minute rotations which normally are so small as to be undetectable by people, but which are enormous deviations of a laser transmitter generator mounted in such a vehicle. As an experiment, try focusing a small-beam flashlight on some object a long way away and then try to run and keep the focus correctly aligned. Highly sensitive and completely automatic tracking systems with laser communications equipment are a necessity. This is in addition to automatic frequency tracking and phase or polarization tracking and generation and power output stability.

Two orthogonal polarization states of a light beam carrier can correspond to the well-known digital states of 1 and 0. In such a system, automatic polarization compensation is provided by applying a dither modulating voltage to a cell which exhibits electro-optic effects. A dither is a voltage which varies rapidly and periodically so as to keep the object it controls in motion to eliminate the viscid friction delay that might occur if one had to stop and start the object each time some intelligence or control signal was present.

The electro-optic cell controls the relative phase of the electric field (polarization) of the input light beam enabling the dither frequency component of the difference of the instantaneous powers in the two polarized signals to be coherently detected. A signal from the detector is fed back to the cell after integration to form a polarization-bias-compensating servo-loop.

A MEDICAL COMMUNICATIONS EXPERIMENT

Some time ago an experiment was carried out to determine if it was possible to convey sufficient information via a laser beam so that a nurse anesthetist located in the V.A. Hospital at Cleveland could obtain instructions and guidance from an anesthesiologiest at Case Western Reserve University located some 1.2 kilometers away. In order to be able to handle this medical situation properly, television signals, voice signals, and data signals on the physiological condition of the patient, such as electrocardiogram, pulse, respiration rate and such, had to be accurately transmitted. This was accomplished in the experiment.

When viewing this medical situation the implications are far beyond just the ability to transmit a lot of data over a laser link such as was used in this case. One realizes that nurses as anesthetists administer a large portion of the anesthesia in medical situation. The ability to be able to provide guidance and help via such data transmission systems that are immune to interference may go far beyond the case of just anesthesia. Visual, voice, and data transmission over the same link simultaneously provides a proper input of accurate diagnosis and medical application.

EARLY PUMPING EXPERIMENTS FOR A NdYAG LASER

An experimental program was conducted to determine and evaluate the alkali lamps for use in a space qualified NdYAG laser system. The alkali lamps had bores from 2 mm to 6 mm. Measurements were made of the lamps' performance both in and out of the laser cavity. One significant observation was that for a constant vapor device that spectral and fluorescent output did not vary for vacuum or argon environments and this led to the use of this laser type in an argon environment. The alkali lamp was finally optimized at a 4 mm bore and for a 430 watt input the laser produced slightly over 3 watts of coherent beam output.

ATMOSPHERIC ATTENUATION OF A LASER BEAM

The attenuation of a laser beam propagated in the atmosphere has been measured. The laser used was a 10.6 micron carbon dioxide (CO_2) laser and the beam strength was measured by a pyroelectric detector after it had traveled one kilometer through the atmosphere. Using a 20 mW power output level, the focus of the target placed at the focal point of a 25 centimeter focal length spherical mirror is one square millimeter. Rain and haze produced no noticeable attenuation but heavy fog gave attenuations of 3,000 times what would be normal without any attenuating materials.

TIME DELAYS IN OPTICAL TRANSMISSION SYSTEMS

Normally, time delays in optical transmission systems are not much of a problem, since the speed of electricty and light propagation and the propagation of microwaves is so fast. But there are some applications in which time delays can be very important. Also, the transmission of pulses (that contain information either singly or in groups, such as a pulse-position modulation system) may be affected by time delay in the system.

Time delays can occur in the equipment used to code and decode such information, in the actual transmission times of the rays of light, and in the response times of detectors and generating times of the laser units. This can be especially true when there is any kind of off-on program operating the devices. There is also the problem of synchronization required in the correct operation of some electronic-optical systems. Synchronization can be adversely affected if the timing is not precisely correct.

MORE ATMOSPHERIC TESTS

Currently the primary type of data transmission is digital. This fits into computer patterns neatly. This is the best way to get desired information from one point to another with the least loss of data and degradation of data. Some digital data transmission tests were conducted using a laser beam propagating through the free atmosphere to determine error frequency and magnitudes. Error frequency was determined using a pulse-code modulater that modulated an argon laser beam propagating with a one watt output and operating on its green line of 514 nm (nanometers). The path was short (3 kilometers) and led over build-up areas. The results were not encouraging. It was shown that this form of digital data transmission is suitable only for specific applications.

SPACE EXPERIMENTS WITH LASER BEAMS AS COMMUNICATIONS-DATA CARRIERS

Having explored the attenuation of laser beams by our own atmosphere and being aware that we are going into space, the next idea is: How can we use lasers in space applications? It is immediately apparent that because there is no atmosphere attenuation in space it should be easy to propagate the laser beams there, so communication to and from satellites is a possibility.

The Aerojet General Corporation conducted some experiments along this line and called them the LCE, laser communications experiments. In the experiments, CO_2 lasers were used both in the vehicle and on the ground to establish communication between a synchronous orbiting satellite (meaning it had to be some 23,000 miles into space above the earth) and a ground station. The problems of beam generation and acquisition by unmanned satellites were also part of this study.

The space environment is good for lasers because the sun is ever present in space, sunlight can be converted into electricity. There is abundant power available for electronic uses once it is converted from photons to "holes and electrons." If the balance of the electronic devices, integrated circuits, and such can be devised in such a manner that they can withstand the rigors of space, then there is an opportunity to put something into space and use it for an indefinite time. How fascinating!

Of course, the Voyagers' missions have run into years, and satellites have been reactivited which had not failed, but were purposefully "shut-down" by their masters. So the concept in a sense has been problem. Now let's think of the possibility of ejection of heavy ions from the propulsion system of some kind of space ship; this may be a system of operation and propulsion which can go anywhere for any length of time, as long as there is sunlight to make possible the creation or conversion of electricity from sunlight. It also requires a new look at the technology that is developing at such a rapid rate. How nice it would be for that same technology to develop sunlight conversion units, which unlike our present, rather inefficient, cells, could operate with an efficiency of 80 to 95 percent. Not only would that technology be acclaimed for the applications to space applications, but also for more mundane applications right here at home.

MODULATING A MICROWAVE CARRIER WITH A LASER BEAM

With a CO_2 type laser, the output might be combined with that of a microwave system in such a manner that the resultant sideband has the results of the sun and differences of the two frequencies. This might be usable in optical image radars and high data-rate systems according to a study by United Aircraft Labs of East Hartford.

The goal of their program was the demonstration of efficient sideband generation of a 10.6 micrometer CO_2 laser carrier at X-band microwave frequencies in a thin film nonlinear GaAs optical modulator element interfaced with a microwave ridgeguide structure. The electro-optical interactions in a GaAs thin film between an optically guided wave and either a traveling wave or a synchronous microwave of the standing type were capable of producing the sum and difference frequencies of the two interfacing signals. A chirped (changing frequency) optical signal was generated by using a frequency modulated microwave signal. This signal can be separated from the others by a proper optical filtering technique such as diffraction grating or a narrow bandpass interferometer.

SOME CONCEPTS OF AN OPTICAL RADAR SYSTEM

The Jet Propulsion Labs of CIT have shown that an optical radar system is feasible. The system is composed of an optical cavity with a laser and mode-locking means. They are used to build up an optical pulse, which is radiation in the optical spectrum.

An optical switch was provided in the cavity to convert the polarization of the optical pulse generated within the cavity. The optical switch was made of an electro-optical crystal driven by a time-delayed driver circuit, triggered by a coincident signal made from the optical pulse signal and a *gating pulse signal. The converted optical pulse signal strikes a polarization sensitive prism and then it is deflected out of the cavity toward the target. It consists of a pulse made up of the optical energy generated by the laser during the pulse build-up period. Then, as in all radar systems, after this pulse strikes the target, some of its energy is reflected back to the starting point and through proper receiving equipment its energy is converted into a kind of signal that can be compared to a timing pulse in a timing circuit to determine the total travel time to the target and back. Thus, by proper interpretation of this time, the range to the target can be determined.

It is interesting that using this kind of radar system, the angular measurements can be much more precise than with conventional types of radars because of the narrow beam transmitted. Microwave radars have a relatively broad beam and thus leave room for a certain region of target-location uncertainty. This error uncertainty is vastly reduced with an optical radar, although acquisition of the target may be much more difficult because of the narrow optical beam generated by a laser.

This brings up the idea of some experimentation conducted by some agencies to try to widen this optical beam for target acquisition purposes. Some have believed that defocusing the optical beam may tend to widen it in all cross-sections, thus making it possible to acquire a target easier than by automatic focusing of the beam.

If this defocusing system is not feasible or practical, then some believe that a microwave radar might always have to be used in conjunction with a thin-beam laser as the acquisition device. This device then points and positions the laser equipment so that it can then track the target, giving acquisition of targets and precision tracking, which is growing more mandatory all the time. Precision tracking is not just for a motionless type of target, but a fast moving and even maneuvering type of target—even if the movements are slow compared to the propagation speeds of an optical light beam. The acronym OPDAR is used to mean Optical Detection and Ranging systems. Notice the similarity with the word acronym RADAR (Radio Detection and Ranging).

THE CONCEPT OF A MULTIPLEX DIGITAL LASER GENERATOR

An interesting concept is multiplexing. This means imposing several channels of information on one carrier in some manner or another. In this case it is a multiplexing of digital information on a laser beam and its general method of operation is fascinating. Time multiplexing, commonly in use on hard wire computer lines and on radio systems, means that some space of time is devoted to each channel of information sequentially. All channels are not transmitted simultaneously. Only in frequency multiplexing as applied to audio and rf signals can the combined output of channels be used to modulate a carrier for concurrent transmission and then decoded by appropriate filters, etc., on the receiving end.

In this laser concept of multiplexing, the frequency components of the laser emission are spatially separated inside the generator cavity. This can be done with a proper prism system. Then each component frequency is separately pulse code modulated according to a digital input signal, transmitted to the receiver in a recombination of the beams and then separated for decoding. The receiver system is relatively simple. If the beams were not recombined and were transmitted as separate beam elements or constituents, then simple photo-to-electric detector systems or infrared-to-electrical detectors might be used for the receiver.

In a gas laser (CO_2 type) what happens is that the gas is caused to pass over some electrodes mounted transversely to the gas flow. Then a spark gap provides added pumping energy when it is discharged into the gas flow causing lasing action. It is said that the gain-bandwidth of an atmospheric CO_2 type of laser discharge is sufficiently wide (around 4 gigahertz) to permit as many as 20 axial modes to oscillate simultaneously within the cavity. Proper filtering can separate a desired mode of operation.

USE OF THE CO_2 LASER IN PRODUCTION

Making printed circuit boards of smaller and smaller sizes is simplified by using a CO_2 laser to drill the appropriate holes for mounting components. The Coherent Everlase CO_2 lasers are used in this operation. A modern circuit board requires many thousands of holes which are very small in diameter (0.004 inch) to fit component lead wires of 0.002 inch. It is almost a requirement that the laser head be controlled by some automatic process. A computer-controlled numerically-specified programmed machine (CNC) performs this positioning readily and accurately. In addition to the CO_2 type of lasers, the NdYAG, and Nd: glass lasers are being used in manufacture to drill, cut, burr, and machine such materials as plastic, rubber, metal, wood, ceramic, and glass. The applications in manufacture include cutting, drilling, engraving, heat-treatment, soldering, welding, and even insulation removal.

During some welding processes, it is necessary to restrict the heat build-up or the material suffers some thermal distortion or other adverse effect. Because a laser beam can be pulsed at a variable rate, the rate can be set to avoid such thermal problems. Thus there is a definite advantage using lasers in some applications over other types of welding or soldering.

The Coherent Industrial Group, 3210 Porter Drive, Palo Alto, CA 94304 has produced an automated laser welder which is a CO_2 type rated at 575 watts continuous power with an x-y moving optics system controlled by a computer-numerically-controlled machine to generate complex welding patterns. The use of optics with the laser becomes essential. This is called a large-scale laser, and it indicates the possibilities of usage for laser machines of this kind. Of course higher powers from lasers are available. Using a laser in a cutting or welding application that requires a high degree of concentration of laser energy means that several factors affect the performance of the unit: the cleanliness of the optical system, the cleanliness of the nozzles, the number of mirrors in the beam path, and the gas pressures if oxygen is added to the cut or weld to facilitate cutting or welding. A good clean cut is also a function of the peak power used.

Because a laser beam can be focused to such a small area, it can be used to solder in very tight spaces. The depth of the heat can be precisely controlled. This also means that in wire stripping operations, the insulation can be removed from cables easily and completely and without any damage to the wires, which might occur if a mechanical wire stripper was used. Another advantage of this type of stripping seems evident if one considers a smaller and smaller diameter wire or wiring to be stripped and soldered. There is a point of diminishing returns when using mechanical devices. Another idea which persists is that the use and operation of a laser system can be modified by a change of optics — not necessarily a change in the whole laser system itself. As we have pointed out previously, once a laser beam has been generated in a proper frequency and in a proper amount of power, control of that beam is required and this is done through the use of the various beam control elements previously described.

ABOUT THE FABRY-PEROT INTERFEROMETER

An interferometer uses the principle of splitting some signal, ray, or other phenomena and then recombining the two halves after each has been subjected in some way to some outside force or forces. In this way one half becomes a reference against which the other half can be compared, and the changes from the original state or condition can be measured, evaluated, and used as a basis to determine the magnitude, phase, or effect of the forces that have been influential in changing the other signal.

In the generation of a laser beam from a diode strip laser element, the ends of the element are manufactured so that they are optically flat and parallel so the photons can stimulate others and cause a lasing process. It is this parallel mirror structure, is called a Fabry-Perot interferometer. It can be constructed on a larger scale by the proper alignment of two first surface mirrors placed so they are exactly parallel. In the actual interferometer mode, the mirrors are used to separate or combine separate frequencies so that the comparison effect stated can be obtained.

A LASER REMOTE-CONTROL SYSTEM CONCEPT

Of course, if the laser beam as a channel can be used for TV, audio and data communications, then why not use it for some kind of remote control application? There are several good reasons why this is a good idea. First there must be the equivalent of a direct wire channel to and from the control site to the controlled object. Second, the channel is relatively immune to outside interference. Third, the equipment does exist for fabricating such a system. You might use any of the conventional "control" code systems over such a link. Only the communications channel is changed, and its equipment is added to an existing type of remote-control system. There is no reason why the number of control channels over such a link would not far exceed what we currently have developed.

What seems to be a problem is the requirement for some kind of line-of-sight with the remotely controlled object. If you use a light beam, which is what a laser beam amounts to, then you must be able to direct that light beam to the target receiver. Thus such a system might not be usable under some out-of-sight situations. Second, haze, fog, and other atmospheric conditions might attenuate the transmitted signal so that inadequate reception takes place. Some of these disadvantages might be removed by control of objects from space. There are some situations where acoustic, radio frequency or hard wire communications are just not possible or feasible for remote-control applications.

The concept of using a very thin solid-state laser radio-control transmitter to control some object from some distance away is appealing and perhaps practical. In some experiments done in this area a helium-neon continuous wave laser was used with a receiver that had a 360° light detector antenna.

MODULATING A LASER BEAM
WITH FM AND AM TYPES OF MODULATION

Acoustical modulation of the amplitude and frequency of a laser beam from a cw helium-neon laser has been accomplished. The fact that this modulation can be accomplished with electroacoustical devices opens a new field of experimentation and development of new types of modulation devices of this kind.

INTEGRATED OPTICAL CIRCUITS

First there were integrated electronic circuits; there are integrated optical circuits. The idea of developing an integrated optical circuit capability for use in a multi-GHz telecommunications application for the military was conceived and put into development. What was desired was a fiber-optic transmission line, multiterminal multiplexing system through low-loss couplings and suitable modulations. New developments will bring about more concrete joining of the technologies of solid-state electronics, fiberoptics, and integrated circuitry.

USING THE PIN DIODE IN A LASER COMMUNICATIONS SYSTEM

An experiment was conducted using lenses and fiber optics so that a curved path could be traversed by a laser beam generated from a pulsed gallium arsenide injection laser when pulsed at about a 10 to 15 kHz rate. The PIN diode used in the receiver is a photo-detector. The method of communication is interesting. It is pulse-position modulation. This requires excellent stability and synchronization. In fact, it is the method of radio-controlling so many models. In this case, each pulse does not represent a control channel and therefore does not cause some control (physical) element to move left, right, or neutral position. The communications concept says that each pulse may follow a sinusoidal variation about its nominal position and thus represent a tone or some date. Each pulse represents a complete communications channel. This is not a new concept but one that has been used for a long time in telemetry applications.

The University of Southern California Electrical Engineering Department has made a study of the Pulse-Position Modulation (PPM) as a basis for a communications system using narrow pulses of light. Synchronization is mandatory and has to be very accurate. The use and presence of signal energy in the form of such optical pulses suggested the use of the pulse-edge tracking technique in a binary PPM system to ensure the necessary tracking and timing. The timing error variances were a function of the system's signal-to-noise ratio. This is not an unexpected result of the study. When you consider that individual channels of separated pulses are designated by their timing with respect to some transmitted reference then it is readily apparent that precise timing must be maintained.

Also, any noise in the system that tends to mask a pulse or reference signal can destroy the synchronization which is the basis for the channeling of the pulses and the amount of excursion of the pulse, which is the data information or communications intelligence.

The Hughes Research Labs made a study of the various modulation methods that might be applied to laser-optics systems. Whole systems were considered from the origin of the intelligence through the transmission over optical carriers to detection and display or creation as a signal usable to humans. Modulation systems such as intensity modulation, and both direct and coherent (hetrodyne) reception of signals was considered. One basis for consideration was the information capacity of such systems and the message signal-to-noise ratios.

These studies and experiments reveal that, as in any communications system, the detection of the signals that contain the intelligence are performed with devices that also detect noise in the phenomena used for transmission. It seems impossible to separate the two completely. But it is the ratio between what can be used and what cannot be used that becomes very important. It is the basis for many studies and acceptance or rejection of various concepts of intelligence transmission.

In the detection of mode locked (fixed-frequency output) laser signals the passive Fabry-Perot cavity was shown to be a good practical approach to the matched filter optimization for the sensitive detection of the mode-locked signal. Doppler measurements of the relative motion over a wide range of velocities of propagation are possible simply by measuring the cavity length for a peaked output.

If the length in a cavity is equal to an integral number of half-wavelengths, the cavity resonates at that frequency. Thus for a long cavity — encompassing an integral number of half-wavelengths for various closely related frequencies —you can see how the cavity might resonate at any of the several closely associated frequencies. Given a long cavity length, as is found in a laser stripe-line generating system, you might have a number of frequencies resonating and separated by some amount of the spectrum. Some stripe-line cavity resonators may be as much as 1,000 wavelengths long, and so there is the possibility that other than the desired frequency will be produced. There may be as many as five to ten modes of operation evident.

AN ANTENNA FOR A 10.6 MICRON PRINTED CIRCUIT

Consider a printed circuit antenna array for a wavelength of 10.6 microns (10.6 millionths of a meter). That is not very long! But, it was found feasible to develop. The analysis was by the method of moments. An infrared dipole was considered. With printed circuit technology, it had to be a two dimensional type of dipole that had both lateral and axial components of current. The subsequent radiation pattern for such a dipole was analyzed and it was found that a pencil beam that was perpendicular (normal) to the plane of the antenna was capable of production. Of course, solid-state optical-guide types of antennas also produce such beams. But, it is interesting to know that considering the radiation theory concept an antenna for the smallest of frequencies could be developed. Going higher and higher into the frequency spectrum, the quantum theory of propagation and analyses is based on this theory. The quantum theory is much used when evaluating propagation through various atmospheric conditions.

POTENTIAL USES OF LASERS

Laser usages in medical and biological applications are increasing almost by the hour. There are also uses in chemical processing and analyses. Initialing controlled thermonuclear reactions can use lasers, and the advantages and desirability of using laser beams for communications are obvious. The broad bandwidths that are available and the security of the systems make this a very important developmental area. There have been studies that suggest the use of laser beams for study of pollutants in detection and also monitoring of pollution. There are countless applications in the geoscience fields, holographic applications, and in manufacturing and related processes. There will be other uses which are not as yet envisioned, but which you might imagine.

SAMPLING A LASER BEAM WITH ACOUSTIC PULSES

There is a patented method for sampling and scanning a laser beam with acoustic pulses. The pulse is propagated across the light beam and the pressure variation (of air) which comprise the pulse tend to form a diffraction grating for deflecting the light beam over its volume of interaction with the pulse pressure waves. As the pulse propagates it doesn't all do so instantaneously; therefore the gradual movement of the pressure wave through the beam causes diffraction to take place gradually. There is a scanning effect produced. If the beam is focused through suitable optics and on a suitable detector, an analysis of the beam in its parts can be undertaken. This is useful in analyses of various chemicals and elements.

A VIDEO-LASER RECORDER

Ampex is working on a two channel 16 mm laser beam film recorder for television sensors or cameras. The two channels of video at 525 or 875 lines per frame are recorded at 30 frames per second with the frames on side-by-side format on 16 mm film. The system has a neon-helium type of laser with its necessary optical system. A 70 percent efficiency was recorded at 16 MHz bandwidth. Here, a TV camera modulates a laser beam, which is coupling its light output to a film record system. Although a great amount of equipment is necessary in such a system, but it is another use for a laser-type light beam.

THE LASER BEAM PHENOMENA IN MODULATION

A trend in laser technology and one that boggles the mind is the fact that acoustic waves can cause an effect which causes a laser beam to change direction (diffraction) or can cause modulation of the light beam. Somehow the combination of acoustic pressure waves with electromagnetic energy that seems to occur in these instances is not obvious.

Within the atmosphere, the air molecules that normally absorb and impede the radiation of light rays or other rays in the higher electromagnetic spectrum, can be influenced by various devices that may cause them to contract and expand. It is not beyond reason to say that if the air changes due to acoustical pressure waves or due to some transducer device, then the light passing through this air, shall be affected. The absorption rate would change. The mixing of the pressure wave front with the electromagnetic wavefront suggests that the beam of rf waves will be refracted or deflected as the case may be. If both the acoustic wave and the light wave are impressed on some transfer device such as a crystal, mixing takes place that causes interference patterns.

THE LASER IS A RATE DETECTOR OF ROTATIONAL SYSTEMS

Strangely enough it is possible to add to the velocity of propagation of a laser beam the very tiny velocity of roll which may be due to the turning of a rocketship, airplane or satellite. You can extract that information to determine the roll-rate of the object precisely. The sensor for such a system of rate detection is a laser interferometer which detects the fringe shifts caused by the body's rotation. Here is a somewhat technical explanation of how the system works: A light emitting-diode pumps an optical-fiber laser (a laser bead at one end of an optical fiber). This sends light rays through the optical fiber or distribution waveguide. The coherent rays from the laser are then partly backward coupled by means of an acoustic-optical grating into a reference optical-waveguide (optical-fiber). The coherent radiation is also coupled into each end of a multi-turn optical waveguide positioned so that it serves as an interferometer. A pair of directional couplers, one for clockwise radiations, extract the coherent radiations from the interferometer fibers after the light rays have made at least one passage around and through the device. Finally, both the clockwise and counterclockwise signals are combined and the phase shift between the two signals is measured. The amount of phase shift is directly proportional to the rotational rate of the system holding the interferometer.

This seems like a hard way to accomplish the rate determination but we need to think of the advantages. First, a gyroscope, which also can measure rate by the rate of its precession, has a tendency to drift so that its axes do not align perfectly with the axes of rotation. Thus an error can be introduced. Second, a gyro wheel has friction. Any variation in its rotational rate affects the precession rate and thus the accuracy of measurement.

Many years ago I examined a rate gyroscope designed by the German scientists at Peenemunde to be used to cut-off fuel flow in the V-2 rocket system. If a high degree of accuracy was to be obtained, it was necessary to constantly determine the velocity of the rocket and shut down the fuel at a precise rocket velocity so that a precise range would result. It was fortunate that such a device did not appear, at least in any numbers, in the rockets launched against London. Instead of the ten-mile errors of the first V-2 rockets, improved types could have hit, directly on. Considering this laser-type interferometer, the accuracy and guidance of weapons of war come to mind. Precise accuracy over millions of miles of space is now possible.

THE YAG LASER

This is the yttrium-aluminum-garnet type of solid-state laser, mentioned by the abbreviation (YAG) in much of the current literature. Some YAG lasers operate at about 1.06 microns in the near infrared portion of the spectrum and they can be used in ranging systems quite effectively

A LASER FLAW-DETECTION SYSTEM

The method used with lasers and opto-acoustics replaces a type of flaw detection system used in industry that uses piezoelectric transducers. A flaw in a system is detected by the transducers that are physically in contact with the surface to be examined. Acoustic waves change phase according to the flaw and thus can be detected.

If you use a pair of lasers and a suitable optical system, it is possible to use one laser to generate a high-powered pulse that is converted by the material into an acoustic wave in the material. The second laser is used as an interferometer that monitors the surface motion and thus the detection of the flaw is apparent by the phase shift as previously described. This system needs no direct contact with the material to be examined, nor is there any limitation on the size or shape of the object under test. The speed of examination is very fast, according to Harry Diamond Labs at Adelphi, MD.

What seems to be almost impossible, yet is very possible, is communications from a satellite to an underwater submarine through the use of underwater acoustics and a scanning laser beam. It seems that the pressure waves from the submarine's acoustic transducers can cause the ripples in the water that normally are so small and fast they are undetectable. Due to the infinitesimally small wavelengths of laser beams and the fact that they may be propagated with a given polarization, such physical disturbances can be detected either through a change in polarization, or phase differences or similar phenomena. This tiny measurements with a laser beam are also much in evidence in a study of geophysics wherein minute changes in the earth can be determined, magnified and studies.

Laser beams can represent the sensing element for such studies.

Lasers also help to align sawmill blades, guide the cutting of marble slabs, and position patients for various medical procedures. In addition the use of laser beams from microscopic size laser generators can, according to recent reports, clean out human arteries and veins by removing fatty deposits or breaking them loose. They can also cauterize human tissue such as the retina in the eyeball, cut away undesired tissue such as cataracts in the eyes, and furnish light sources for inspection inside the human body. This chapter has already mentioned drilling holes, and welding, soldering, and cutting usages in industry.

Chapter 4

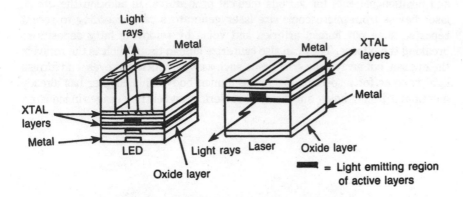

Light rays

Metal

Metal

XTAL layers

XTAL layers

Metal

Metal

LED

Light rays **Laser**

Oxide layer

Oxide layer

■ = Light emitting region
of active layers

Some Required
Background Information
and Concepts

ALMOST EVERYONE WHO HAS HEARD OF FIBEROPTICS KNOWS THAT IT IS SOME
kind of a waveguide. It is a flexible strand of some glass that is able to
channel light rays through it in an efficient manner.

The fiberoptics channel does not permit the desired light rays to escape
from its interior. It is capable of handling two-way transmission of light rays in
such a manner that one can (but must not) look into an end and see what is
happening at the other end. This is only true in some scientific and medical
applications. Also, because there are ways of transmitting sound and pictures
over these light rays, fiberoptics transmission systems open up a completely
new concept in communication systems with unique advantages with regard to
noise elimination and interference.

Light rays are similar to radar waves which can be deduced from a basic
equation:

$$V = f\lambda$$

$$V = 3 \times 10^8 \text{ meters per second}$$

Since the unit Hertz = 1/seconds, then:

$$\text{Meters/Second} = 1/\text{Second} \times \text{Meters/1}$$

Having matched units, the equation can be restated in the form:

$$\lambda = V/f$$

V = velocity of light, f = frequency in hertz, and λ = wavelength in meters.

As (f) approaches infinity, the wavelength approaches zero. Supposedly that is the end of the spectrum, although there may be other emanations of some different kind than oscillations which occur beyond the oscillatory spectrum. Radar waves are frequencies below the visible light portion of the spectrum, but they are oscillations. Since we are extending that spectrum, then we assume light rays to be oscillating electromagnetic radiations also (but of a much shorter wavelength). Some scientists refer to the corpuscular nature of light and energy parcels called photons.

Light rays are electromagnetic in form, such radiations can be guided inside specially sized and shaped waveguides. Why not use optical fiber as the guide? Radar waves flow through their metallic waveguides because of reflection from the walls and the peculiar manner in which they impart the walls. It is easy to accept the concept that light rays may also be similarly reflected in a different type of substance more suitable to the reflection of light than a metallic wall might be. Light also can be caused to flow down the lengths of channeling forming an optical-fiber path.

The essence of fiberoptics is when light rays are channeled through optical fibers by generating the proper frequency of coherent light permitting easy passage through the channel material. This implies that different types of substances (all of which fall into the optical or glass category) have some different characteristics that make the conductivity of certain light frequencies easier than other light frequencies. In an elementary concept it simply means some types of optical fibers conduct some light frequencies easier than they do other light frequencies. We must generate the proper light frequency to pass through a given type of fiber, just as we have to generate a certain radar frequency to pass easily through a waveguide of a certain dimension.

MODERN CONCEPTS OF LIGHT AND LIGHT PRODUCTION

The creation of light is associated with the passage of electricity through a wire in a glass bulb (our ordinary tungsten filament light bulb) or the passage of high potential electricity through such a gas as argon in a neon tube. You also know, from elementary physics, that light is a component of the electromagnetic spectrum. James Clerk Maxwell showed this to be true by using Gauss's law for electricity and magnetism, and using Ampere's law in an extended form and Faraday's law of induction. He showed that the light rays were waves which are electromagnetic in nature. He also showed that all such rays or waves have the same speed in free space. We know that speed as 3×10^8 meters/sec. The range of electromagnetic radiations extends from a zero frequency (hertz/sec) to an infinite frequency, and those of which light is a section consist of frequencies whose wavelengths in λ centimeters range from

10^{-2} through the ultraviolet region of 10^{-8} on up in frequency through the gamma-ray region of 10^{-14} wavelength.

The speed of light is fairly constant over infinite distances. Over short distances however, the speed of light varies as it passes through different conducting media. So long as the total speed component over an infinite distance in free space remains constant, we have not violated the law which Maxwell derived in his classic work. Experiments have shown these finite speed variations to be a reality and from these conditions many scientific developments have evolved such as lenses and filters. Since the days of Galileo, who was concerned with the speed of light question, values have been measured by various means. A value of c, the speed of light in a microwave cavity, has been determined to be 2.997925×10^8 meters/second, whereas Galileo found by his measuring means a speed of light to be 2.997924×10^8 meters/sec.

Here are some more definitions:

- ☐ Micron $= \mu = 10^{-6}$ meter
- ☐ Millimicron $= m\mu = 10^{-9}$ meter
- ☐ Angstrom $= Å = 10^{-10}$ meter
- ☐ Visible light spectrum = approx. 450 – 675 ($m\mu$) (to the human eye)
- ☐ Center of visible light spectrum approx. 5500 Å
- ☐ Reflected energy from absorptive surface:

$$P = \mu/c = \text{momentum energy/speed of light}$$

- ☐ Reflected energy from totally reflecting surface:

$$P = 2\mu/c = (2) \text{ momentum energy/speed of light}$$

Wavelengths of light are measured in extremely small distances, the Angstrom being:

$$1/10,000,000,000$$

of a meter. You can calculate the length if you know that a meter is 39.370078 inches. It is difficult to imagine such a small distance which is representative of one wavelength of light. Light fibers or light conductors have to be designed in terms of the wavelengths of light that they are to be able to channel.

A point of interest to us is to know that the orange-red light of krypton-86 has the most sharply defined wavelength and is used as a standard of measurement. Length measurements using a green light from a mercury-198 lamp is said to be accurate to one part in one billion! Our modern electronic instruments which measure frequencies and wavelengths of light enable us to make and use such small values to the above accuracy!

The production of light is sometimes accomplished by heat. Witness the heat generated by an electric light bulb when electricity passes through its filament. Light may be generated by lasers, when the atoms are stimulated and caused to go from one energy level to another by the addition of energy to a synthetic ruby crystal or some kind of appropriate gas. Energy from a light bulb is noncoherent. That is, the waves strike out in a spherical or random manner from the filament much like the radio waves or microwaves emit from a suitable antenna for those frequencies. Lasers, on the other hand, emit coherent light rays. This means the wave fronts are parallel. The frequency of the light being generated is a function of the type of laser that is the source of such light.

A source for light is needed to send that light beam down some kind of special conducting fiber with as little loss as possible. Some types of material conduct some frequencies of light better than they conduct other frequencies, so the frequency of the generated light is important to us. Also, whether or not the light is coherent may have some effect on its propagation through such a fiberoptics channel. We can begin to determine the correctness of our thinking by investigating light reflection and refraction. (The book Modern RADAR (TAB Book number 1155) is a source for more information about radar frequencies and waveguides and related ideas.)

LIGHT REFLECTIONS AND REFRACTIONS

If a light beam goes from one medium into another optically different medium, it is bent. One familiar example is when a beam of light goes from air into and through some water as illustrated in Fig. 4-1. Notice that the ray of light does not continue in the line of the air ray. It bends through some angle theta (θ) as shown. It has been shown in other works how a fish sees a human where he is not physically present. Or how a human looking at a fish in the water sees that fish in a spot where he is not really located.

It is important to consider how the ray is described as it approaches some different medium surface. The angle, measured from a perpendicular to that surface, is called the angle of incidence. There is an axiom in physics that light and other type rays will be reflected at an angle equal to the angle of incidence (provided that the angle is not so great that no reflection takes place). Figure 4-2 shows a section of rf waveguide and define these angles as they apply to the transmission of microwaves through that waveguide.

A critical angle of incidence (such that to exceed this angle or not to exceed this angle) might result in a nonreflection situation. If you have studied the transmission of radiowaves, then you may recall when the waves hit the Heaviside Layer, some are reflected and some are not. Those which are not reflected hit this layer at less than the critical angle for reflection and so pass through the Heaviside layer along a refracted line.

THE LAWS OF REFLECTION AND REFRACTION

As in all good physics texts, the laws for reflection and refraction can be stated as in Figs. 4-1 and 4-2.

(A) The reflected and refracted rays will lie in a place containing the

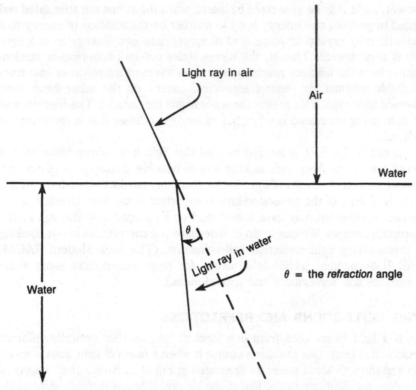

Fig. 4-1. The bending of a beam of light at the interface of air and water.

incident ray, the perpendicular (normal) to the surface of the material of reflection or refraction, and the angle of incidence will equal the angle of reflection.

(B) The equation: sin of angle of incidence/sin of angle of refraction $= \dfrac{n_2}{n_1}$

where n_1 and n_2 are numbers called the indices of refraction.

The air-water index is about 1.33, crown glass-air is 1.52 and quartz is fused to have an index of 1.46. The air medium for these figures is assumed to be vacuum-like in composition with a refractive index of (1). These index figures are for a given wavelength (5980 Å) and may change for other frequencies.

These are some basic principles which are important when considering fiberoptics. In order to get a reflection there must be a junction of two types of material such that there is some index of refraction between them, and the rays must not approach the critical angle or they will pass through and not be reflected at this junction. Prisms are based on the refraction of rays of light. Because different angles of refraction occur for different frequencies, you can use a prism to separate the light spectrum and spread it out on a surface so we can see it. It is also interesting that the laws of reflection and refraction can be derived from Maxwell's equations (if you want a mathematical challenge).

Figure 4-2 indicates a rigid waveguide. In fiberoptics that use flexible fibers there is not necessarily a rigid guide or channel. External pressures on the optical fiber may cause us some problems.

A SLIGHT BIT OF HISTORY

Light has been used as a medium by which some intelligence can be transmitted from one location to another. Early Indians used light reflections from mirrors to signal one another that wagon trains or soldiers were near. In modern warfare, the signal light employed long and short blips with Morse coding between ships to convey intelligence. The advantage is that the countermeasures against such a signaling system are rare and difficult to employ. Covert eavesdropping on such signals is difficult because of the directivity of the light beam. The ability to convey intelligence of the type required is very high. However, the range is very limited as compared to that obtainable with radio waves due to the absorption and attenuation of the light energy by particles, moisture, and such in the atmosphere.

Back in the 1880's Alexander Graham Bell spoke over a light beam. He used a reflector and a lens to focus sunlight onto a device which vibrated

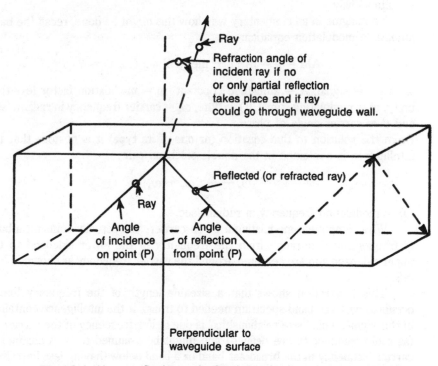

Fig. 4-2. Incidence, reflection, and refraction angles in a microwave waveguide.

sympathetically with human speech. The light beam was made to fluctuate in and out of focus in such a manner that its strength upon a selenium detector was caused to vary. This fluctuation caused a varying electrical signal from the detector which could then be made to activate a telephone receiver in the usual manner and thus re-create the original speech. The distance between transmitter and receiver was very short compared to our modern capability of transmitting light beams. The principle involved was valid and is still a basis for our current communication technology using light as the conducting medium. The method of varying the intensity (or modulating the light beam) is now different using solid-state technology and our detectors are improved, but the basic concept is still the same.

The manner in which light beams are modulated to convey intelligence makes use of the principles of modulation found in modulating radio waves. This results in a much more sophisticated system of intelligence transmission than was ever imagined by Alexander Graham Bell. The light spectrum is a continuation of the radio spectrum and the frequencies increase as the wavelengths become shorter. It has been proven by scientists that the information carrying capacity of a channel such as this increase as the frequency of the channel increases. A multitude of channels of information can be sent over one light beam. By using only a small fraction of the bandwidth available when using infrared or visible light frequencies, one such channel can carry and convey all the telephone conversations of every person on the North American Continent simultaneously!

To imagine in an elementary way how this might be done, recall the basic amplitude modulation equation:

$$AM = e = [A\,(1 + me^s\,(t))]\,\sin\,(\omega_c\,t) = \theta$$

where e = signal, A = amplitude constant, m = modulation factor less than unity, e^s = modulating signal, (t) = time, ω_c = carrier frequency in radians/sec, and θ = carrier angle of phase.

From the solution to this equation (or one of its type) it is obvious that the intelligence is contained in the double sideband pair:

$$\omega_c - \omega_s;\ \omega_c + \omega_s$$

ω_s = modulation frequency in radians/sec.

When amplitude-modulating a given carrier, you actually transmit a band of frequencies that range from the frequency of the lower sideband to the upper sideband and are centered around the carrier frequency as shown in Fig. 4-3.

This illustration shows that a sizeable length of the frequency line is occupied with the band spectrum needed to transmit the intelligence contained in this signal. This "size relationship" is due to low frequency of the carrier ω_c (as radio frequencies are defined) and might be assumed to be a channel or carrier frequency in the broadcast band or a band below (having less hertz/sec) than found in the broadcast band.

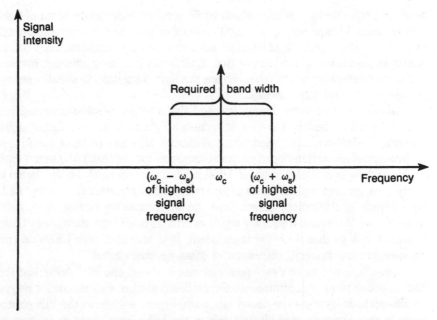

Fig. 4-3. Amplitude modulation (AM) with a radio carrier illustrating the bandwidth required.

Long ago our inventive telephone companies came up with a system of transmission of telephone messages called carrier telephone. Prior to this time the telephone companies had to have one pair of wires for each telephone installed (unless of course several in a home or office were parallel-wired to that single connecting pair of wires). They determined that it was economically unsound to string so many cables containing so many pairs of wires. This was true even if they used a common wire to all other wires to make up the telephone pairs.

In the carrier telephone system, one pair of wires could be used to convey a broadband higher frequency signal. This could be broken down into bands that could be then modulated at one end by the transmitter and demodulated at the receiving end so that a multitude of telephone conversations were then carried over a single pair of wires. It was much more economical with respect to both equipment, installation and maintenance and hence consumer bills were less. The telephone company still made a good profit and everyone was happy. As the population grew, more of the problems encountered earlier began to reappear. Again the telephone company faced the situation and tacked the problem to try to find a solution. That solution will now come about through the use of light-conducting fibers to replace the wires previously used. They can get so many bands in the light frequency range around one broadband light frequency that they can handle virtually millions of telephone conversations simultaneously over one channel.

This development was not without some problems. First they had to develop a suitable light source which could send an appropriate beam of light (at the desired frequency) over a light channel or fiber that was strong enough to be usable in a multitude of installations and over long distances. They had to invent some means of modulating that light beam in a most efficient manner and invent detectors which could reduce the light variations to suitably recognizable voices and other sounds.

Of course, you have already guessed that the laser—first invented and demonstrated in the 1960's—would furnish a light source. Also, LEDs (light-emitting diodes) were invented with sufficient brightness to be of use in this application. Neodymium-doped yttrium-aluminum-garnet (Nd:YAG) lasers emit a very narrow laser beam at about 1060 nm (nanometer) and this line-beam is easily sent through optical fibers of an appropriate type with low losses. LED type devices of the indium-gallium-arsenide phosphides operate at about 1200 to 1500 nm. When this frequency signal is sent through high-silica fibers there is even less loss than with the laser beam. It is said that with LEDs on this wavelength, the material dispersion of glass becomes zero!

Detectors had to be developed, and one of these, the PIN diode, had the fast response time to accommodate the modulation rate. PIN stands for p-type silicon-intrinsic layer-n-type silicon sandwich. Figure 4-4 shows the PIN photo-diode in cross section, and illustrates how the PIN diode works in reverse to that of a light source. The light rays that are absorbed produce the electron-hole pairs that result in a current flow. This current is directly proportional to the quantity and intensity of the light received.

A graph which illustrates the efficiency of some types of PIN photodiodes made from silicon and germanium, as well as from some composite materials much written about in current literature, gallium (Ga) indium (In) and arsenic (As) is shown in Fig. 4-5. Germanium is often used in small solid-state radios as detectors Figure 4-5. This graph illustrates how efficiently a PIN diode made from the stated materials converts the light (at the various frequencies) to electricity.

Figure 4-5 shows that silicon PIN diodes work well at roughly 600 to 1000 nanometers (nm), while germanium and GaAs work well at the higher frequencies of 1600 to 1800 (nm). Notice that the composition photodiode

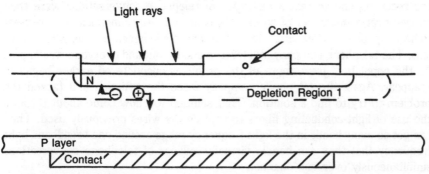

Fig. 4-4. A cross-sectional view of a PIN diode (courtesy of Northern Telecom).

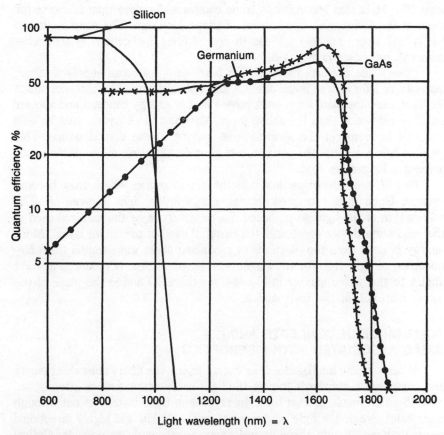

Fig. 4-5. Some relative quantum-efficiency curves for some types of PIN diodes (courtesy of Northern Telecom).

made of GaAs has a nice linear output curve section at the lower frequencies with an efficiency of almost 50 percent. Quantum efficiency is defined as the fractional number of photons incident on the detector that are converted to signal current carrying electron-hole pairs. When the photon energy drops below the bandgap energy of the material the material becomes transparent to the light rays. The germanium detectors have a rather large amount of thermal and shot noise. This is also true of avalanche (multiplying) detectors. Silicon has a low noise level. The avalance photodiode acronym is (APO) and because of its low noise contribution to the current conversion, silicon is used in this application which results in a gain or multiplication of the light-to-current energy.

Bandwidth and recovery time is important in any electronic application and especially in a fiberoptics application. The speed of operation of a circuit and its bandwidth is governed to a very large extent by the RC time constants in the circuit (also the post detection filters used). In most PIN and APD diodes the limit is less than one nanosecond (ns); this in turn ensures a bandwidth of at least 350 MHz that according to some companies is more than adequate for most of the foreseeable applications of these detectors. The construction of LEDs and lasers suitable for use in optical fiber applications is somewhat different, as shown in Fig. 4-6.

Lasers are preferred over LEDs because a laser can supply a larger quantity of light power more directly into optical fibers. The material which forms the confinement areas must have a higher energy bandgap and a lower index of refraction than the active layer. Materials in all layers must be well matched in terms of the atom-to-atom spacing in the crystal lattice. This ensures few defects in the junction region, makes the unit more efficient, and gives it a longer life span.

In Fig. 4-6, electrons flow into the active region where they become excited. Upon their return to normal energy levels, light is given off. The wavelength of the light so produced can be governed by the material used in the solid-state device (especially the material used in the active layer). More energy is used from the electrons in producing short wavelengths (high frequencies) of light than for the longer wavelengths. Light is produced at right angles to the active surface in Fig. 4-6 for the LED and in the plane of the active material for the laser device.

SOME MODERN CONCEPTS AND
IDEAS ASSOCIATED WITH FIBEROPTICS

When sending intelligence over optical fibers, the fibers must meet purity and composition standards to pass the light easily without much attenuation. The fibers are made so that the light rays down them instead of out through their sides. Next, the light source must emit a bright and highly directional beam that gets it into these optical fibers as efficiently as possible. Optical fibers are very small in diameter (a fraction of a millimeter) and must have a composition such that the light-losses are less than 20 dB/km. Many current types have losses much less than this figure. The fibers are made by Corning Glass Works by synthesizing silica glass. The raw materials are vaporized and

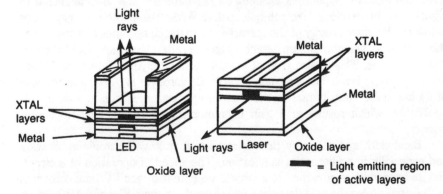

Fig. 4-6. LEDs and lasers for optical fiber applications (courtesy of Northern Telecom).

deposited inside a length of quartz glass tubing which was then drawn into a small-diameter rod.

The modern way to convey intelligence over a light-carrying transmission system is by using the binary coding system or digital transmission system. The source must be capable of being turned off and on very rapidly; the faster this can be done the more intelligence we can transmit. The receivers must be able to receive these light pulses accurately and develop electrical currents from them. In the pulsing process they must recover very rapidly from the impact of each light pulse. A sharp turn-on and turn-off of the receiving element is required to produce the pulse shapes needed for positive indentification, and thus eliminate noise and any other spurious signals which might enter into the system.

You may be wondering what about the use of ruby-rod type lasers or gas lasers as a source of light for fiber optical systems. The ruby-rod may be cumbersome, and the gas laser may have a short life (it is also pretty cumbersome for this type application). Therefore, they aren't appropriate in this application. The LED and solid-state lasers (as shown in Fig. 4-6) are the most applicable to this means of intelligence transmission.

With the transmission channel (the optical fiber) available in the required pure state and strength, and the light source in LEDs or solid-state lasers, requirements for the photodiode became apparent. PIN type diodes and the APD (avalanche) type diodes do the job. They have the necessary conversion efficiency (light to electricity), they are small, and they have the quick recovery time required.

SOME SAFETY TIPS

If you plan experimentation with light producing sources that you may find applicable to optical fiber type systems, be careful! Do not look directly into the light-emitting region of the unit. It emits rays which are very harmful to your eyes and permanent eye damage could result. So be careful and be safe! You can check the output of such devices with a safe system that won't present any danger to you. Learn about this method and use it!

HOW STRONG IS AN OPTICAL FIBER?

The theoretical strength of silica glass (tensile strength) is as high as one million psi! Tensile breaks have occurred in laboratory tests at Northern Telecom at about 5 kilograms of force. Since 2.2 pounds approximates a kilogram, this means a pull of about 11 pounds (or slightly more). But, it is common at Northern Telecom to test such fibers to a level of 300 to 500 grams (one kilogram equals 1,000 grams) to weed out any weak fibers. Sometimes they appear with cracks or other imperfections which would later cause problems if used as they are produced. So, fibers must be chosen after tests and measurements to ensure that they have the correct properties of low-loss light transmission and the necessary tensile strength to stand up in installation and use. The testing used with the fibers has been very exacting high temperatures, humidity, water immersion, and exposure to acids and other types of solutions they might encounter. It is now expected that their life will be equal to that of the well-known copper cable lines now in use in telephone systems. The fibers are so flexible that you can wind them around one finger. During installation care is usually taken to avoid sharp bends and to avoid bending splices so as to minimize the stresses that might cause premature failures.

MULTICHANNEL TRANSMISSION

We have discussed some history in the transmission of multiple carriers, either modulated with sound, TV, or other intelligence (when such information is conveyed by amplitude type of modulation as mathematically stated in Equation 4-4). Now let's examine a diagram which illustrates how such a multiplicity of channels might be transmitted over a fiberoptics system using similar techniques used for carrier telephone systems. (See Fig. 4-7).

In Fig. 4-7B, the channels are actually well separated by guard bands. Because of the carrier frequencies used, the addition of the sidebands does not widen the channel much from that required by the carrier alone. Thus many channels can be transmitted over the same wire pair or through the same waveguide (or the same optical fiber). In optical fiber transmission, the frequencies are very high so there can be a multitude of non-interfering channels. Practically all the telephone conversations on the North American continent could be transmitted over this kind of system at visual light frequencies without having one cause any overlap or interference with another channel.

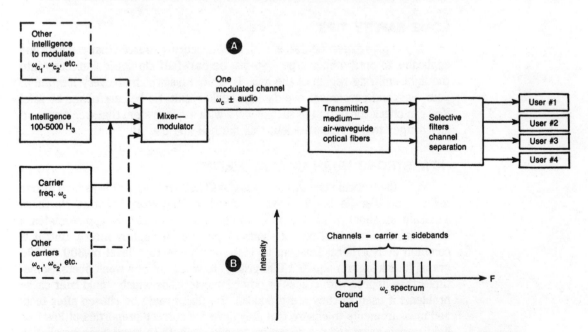

Fig. 4-7. Basic block diagram of a carrier system (A). Carrier transmission of multiple channels (B).

The selective filters at the decoding end which separate the channels are of great importance. Some thought has to be given to the various carriers so that intermodulation (mixing of the carrier frequencies to produce intelligence sideband frequencies which distort the original sidebands) does not occur. There is a system using this type of technique which is called suppressed carrier transmission. Once the carrier has been modulated with the intelligence, the carrier itself conveys no information in an AM system and can be removed by appropriate filtering. This is single-sideband (SSB) transmission much used in current amateur and commercial radio transmission applications. If the carrier is eliminated at the transmitting end it has to be reintroduced at the receiving end for demodulation as is done in current SSB systems. Because the spacing between channels can be reduced, the number of useful information transmission bands is increased.

The carrier system of transmission of intelligence has long been known and used. Generation of frequencies in the visible light region is a fairly new accomplishment, but the production of sufficiently pure optical fibers with the methods of bending them, connecting them to input and output (I/O) devices, and splicing them is a very new development. Like many things scientific, the principles were known or suspected for a long time but the physical implementation of those principles had to wait upon technological development of devices before such concepts could be put to practical use. Such is the basic story with optical fiber transmission of intelligence.

LET US EXAMINE AN OPTICAL FIBER MORE CLOSELY

What is so great about the optical fiber? It is a piece of glass which permits the passage of light, right? Wrong! It is not just a piece of glass. It is a very fine strand of a very special glass which may be only 125 microns in diameter! (A micron is one-millionth of a meter.) It is a glass strand that is about the same thickness as a human hair.

The electromagnetic waves that constitute light tend to travel through a region that has a high refractive index. So, the center of the glass strand — the core — is that type of material. Some glass fibers have a core diameter of only 50 microns and this core has a graded type of refractive index. The importance of this grading is to make the core a little more broadband than a core would be if it was the same index of refraction throughout.

In order to keep the light inside that core, it must be covered with some kind of material with a different index of refraction, or the necessary reflections from the junction of the two materials won't take place. Another coating is formed on the core that is called the cladding which has a lower index of refraction than the core itself. Finally, to make it all look professional, make it stronger, and prevent damage to the cladding, it is common to form a jacket over the cladding which is usually some type of plastic material.

The story doesn't end here. The transmission of digital light-pulses at very high rates through this fiber must be considered. How can multiple conversations, pictures, and so on might be sent over these optical fibers simultaneously? How can light be coupled into such a fiber? What type of connecting devices are necessary to get the light out of the optical fibers?

WHY IS SNELL'S LAW IMPORTANT?

We've been discussing reflection and refraction. Willebrord Snell was the man who discovered the law concerning this phenomena in 1621. What he came up with is an equation which has lasted for the past several hundred years and is important to our concept of how light rays travel down our light tube. The equation is commonly stated as:

$$\text{Snell's Law} = n_1 \sin \theta = n_2 \sin \phi \text{ or } \frac{n_1}{n_2} = \frac{\sin \phi}{\sin \theta}$$

which says that the ratio of the refractive indices is equal to the ratio of the sines of the angle of incidence and the angle of refraction. Since this is somewhat difficult to visualize mentally, look at Fig. 4-8. A light ray coming through the cladding into the fiber-optic core material is bent less than the angle of incidence. Or, to state it another way, the angle of incidence is greater than the angle of refraction.

Figure 4-9 shows the significance of this concept.

Turn the whole thing around so that you now have a ray originating (somehow) inside the core material. It goes upward toward the cladding through an angle that we have shown to be 45.58 degrees (even if it does not measure this with your protractor). Assume it is this value. This is called the

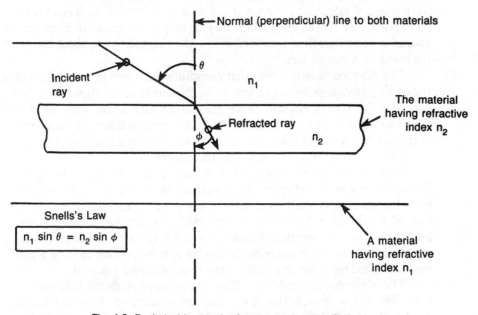

Fig. 4-8. Basic incident and refractive angles specified.

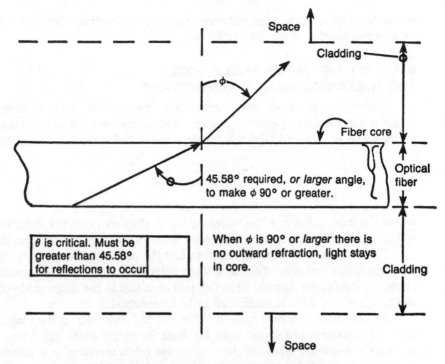

When φ is 90° or larger there is no outward refraction, light stays in core.

θ is critical. Must be greater than 45.58° for reflections to occur

45.58° required, or larger angle, to make φ 90° or greater.

Fig. 4-9. Illustrating the critical minimum-angle.

Brewster angle, or the critical angle and this is important because if the angle is any less than this value some light goes through the core into the cladding and escapes from the light tube. Notice what happens when the angle becomes less than critical. A more direct ray goes upward (as shown). Instead of grazing the boundary between the core and the cladding, the shaft of light passes through that boundary and goes from the core into the cladding area and thence may pass outward into space depending upon the angle at which that ray strikes the boundary between the cladding and space!

The angles are measured from the normal line to the ray. Thus if θ has a value of 90°, it is parallel to the side of the core, or it remains inside the core for this value or a greater value.

DO DIFFERENT MATERIALS
HAVE DIFFERENT INDICES OF REFRACTION?

Different materials have different indices of refraction. For example, if a reference material is empty space, it is said to have an index of refraction of (1). Water has an index of refraction of 1.33, and crown glass has an index of refraction of from 1.41 to 1.61 depending upon its composition and manufacture. A diamond has an index of refraction of about 2.4. The various facets or polished flats on a diamond are specifically angled to help produce the greatest outgoing reflection of light rays which are refracted internally from these flats unless they are already going outward in a desired direction. Thus there are various magnitudes of diamond brilliance.

WHAT DO THE TERMS SINGLE MODE AND MULTIMODE MEAN IN FIBEROPTICS?

A fiber is single-mode when certain relationships hold. Actually, what is meant is that the fiber operates in a single mode. It does not have a mode based on its shape or material. The relationships are stated in equation form as

$$\frac{2\pi a}{\lambda}\sqrt{2n\,\Delta n} \leq 2.41$$

where the core radius is a, the wavelength is λ, the core reference index is n, and Δ is the difference in refractive index between the core and the cladding.

The equation can be manipulated to find the permissible size of the core for a given size and type of cladding if the difference in the core and cladding refractive indices are known. What this tells us is that in the single mode only one frequency of light is considered to be transmitted.

An optical fiber is multimode if either the core diameter or the cladding index of refraction are larger than the limit for single mode operation. In multimode operation many different light rays (each traveling at a different angle of reflection but at less than the critical angle) are guided down and along the core. In the equation stated, it is assumed that the index of refraction of the core and cladding are uniform and the change in the refractive index at the boundary of the two is abrupt. It is possible to have a graded material which gives a gradual change in the index of refraction from the center outward. This decreases the mode dispersion along multimode light channeling fibers.

LOSSES IN OPTICAL FIBER SYSTEMS

When you put light energy into an optical fiber (and do it in the proper manner so that it is transmitted by reflections down the core or along the core material) you find that inevitably the light energy weakens as it travels over any distance. This is due to the attenuation of the signal by the fiber material which absorbs some of the energy in the passage of the light rays. It becomes very important to consider just how much loss can be tolerated in the various types of systems which might use optical fibers for the transmission of intelligence.

In some automobiles, optical fibers are used to convey a small amount of light from the taillights or the headlights to small panels in the front or at the rear of the inside of the car. By glancing at these panels you can see whether your lights are on and functioning properly or not. There are other applications in airplanes, ships, computers, medical instruments, and buildings where optical fibers are used for signaling, intelligence transmission, and even control applications. The losses are tolerable in these short-range applications. The studies show that losses of up to 50 to 100 decibels per kilometer may be permissible. In applications such as in telephone systems where very long runs of optical fibers are used (distances up to 14 kilometers and more), the efficiency of the fibers to transmit light rays must be better with losses equal to or less than 5 decibels per kilometer. Some types of fibers currently in use have losses of only 2 to 3 decibels per kilometer when the light frequency is around 800 to 900 nm (nanometer). This is called the short wavelength region by those industries engaged in developing this means of communications.

In long span arrangements or systems the light might be reinforced at various stations along the way. One would not want to have so much reinforcement (so many repeater stations) that the cost becomes prohibitive when optical fibers are used, because industry is interested in cost effective systems that show a profit.

Now that we have brought up the subject of frequency and wavelength along with attenuation of the signals, let's expand on this idea by saying that some frequencies are attenuated more than others in a given type of fiber. There are many factors related to optical fibers, such as strength, reproducibility, index of refraction, losses, costs, and repairability. One important item is maintaining the transmitting frequency at the proper value to match the optical-fiber system being used. Some lasers are affected by temperatures but they can generate the required energy. Lasers are highly directional. LEDs can generate about as much power when properly excited and fed strong enough currents but have a slower response time than lasers. It is said that the complexity of the laser system is much greater than that of the LED light excitation system. Reliability of the source is also important. The lifetime of each type of device is becoming an area of study. It is said that lasers can have lifetimes of 10^5 hours; LEDs may exceed this by some 10^2 hours of operation. The losses in the optical fibers are from Rayleigh scattering at the shorter wavelengths and by infrared absorption at the longer wavelengths. Rayleigh scattering tends to diffuse the light so that its intensity is not that of a single point source and thus it is weakened.

SOME IMAGINED APPLICATIONS OF FIBEROPTIC SYSTEMS

Let's examine some of the possible and projected and imagined uses for optical fiber systems according to some very inventive and imaginative (and sometimes incorrect!) science writers. One science reporter imagines that it might be possible in the future to send an optical signal to your home from your car to cause the oven to turn on so dinner is ready for you when you arrive. Or, perhaps you want to turn the house lights on or off. You may be able to do it from your car wherever you are. It is said that a telecontrol system which uses optical fibers is operating in the western part of Tokyo. Being a part of the telephone-telegraph system, it reaches into homes and businesses and all such places where special signals might cause selected machines or devices to operate. The regular wires and cables of the telephone exchanges will one day be replaced by the infinitely smaller optical fiber lines. These can transmit even more information, and control signals and pictures more easily and accurately with less problems due to interference. There will be a fiberoptics revolution.

This means an expanded operation from your home, car, or office to do many things that might currently require either your presence or communication by the slower and more uncertain means used at present. Imagine getting your bills before the ink is dry on the purchase request! But don't worry because at the same moment your bank will be extracting from your account a payment and depositing it in the seller's account. If anything makes you ill you could simply connect a bunch of gadgets to yourself and send all your symptoms to your doctor whose computer will tell you what to do about them!

Another report states that optical fibers are appearing in conjunction with new sensing devices. These devices range from gyroscopes that sense inertial references to heat sensing devices of various types and kinds. One type even senses the temperature and operation of a car's engine and then flashes on its LED-type display panel the delightful information that you'd better stop and do something about some problem.

Sensors have been developed to measure strain and stress in materials from airplane parts to buildings (but not in humans, unfortunately). Optical fibers can be used to convey information about the inner workings of nuclear reactors from the inside to the outside of such devices. Such fibers convey light into and pictures out of humans, in medical applications and that the oil industry is using such devices with appropriate sensors to find that black liquid gold.

Because these fibers may be sensitive to changes in stress and strain, temperature, rotation, pressure, and perhaps even sound waves, they may become a part of the sensing systems of the future. All that has to be done is to know what form the light rays have when entering the fiber and then be able to determine what changes have taken place as these rays go through the fiber. When you know how the fiber changes with an increase or decrease in temperature, then you can easily (with a computer) calculate the changes to the light rays and thus have a means of measuring and sensing temperature changes and evaluating them.

Sometimes it is necessary to split the light rays into two sections. This is true in sensing rotation. Let's modify the speed of light transmission by adding the speed of rotation of an object by using a device that rotates in the direction the body is rotating. Such a device might be a long fiberoptics tube. When the original frequency of the light beam is compared with the half that was split off and sent through the fibers rotating, there is a slight difference in frequency proportional to the speed of rotation of the fiber container or housing. So, you can measure the speed of rotation of an object. This application is a new type of gyroscopic sensor and can be applied to almost every type of moving object.

The ability of the fiberoptic transmission system to reject interference from electromagnetic sources can be very meaningful. That means greater reliability in control and in communications. In control applications slight static pick-up of electrical impulses can cause problems in the movement of machine arms, in the control of electricity, in power plants, and so on. This means of communication between electrical and electromechanical devices becomes of great importance in our future.

No one is certain just what the future of optical fibers is going to be but everyone is certain of one aspect of this technology: it will cause a revolution in the electrical, electronics, communication, and computing fields. Its applications are growing at a rate which almost surpasses that of the solid-state integrated-circuit field.

In the field of robotics, optical fibers are being developed to give robots the ability to "see." Optical fiber scanners can be made up of a large number of individual fibers which can be only 0.003 inch in diameter (approximately 0.07 mm in diameter). These fibers are each fitted with appropriate I/O (input-output) connectors to get the light rays of a suitable frequency into and out of the fiber light-tube.

These bundles of fibers can be arranged in a couple of different ways. One is to produce coherent rays of light by arranging each fiber with the same physical orientation of all other fibers in that bundle. This is the type of arrangement commonly used in a fiberscope, which is flexible in construction. The opposite (noncoherent) fiber bundle has fibers arranged in a random manner without regard for the orientation of the light rays or wavefronts emerging therefrom. Recall that if a laser is used as the light source, it provides a coherent type of light imagined to be a light ray with all wavefronts parallel in a given orientation. If the fiber is now twisted, that wavefront will twist with the fiber. It maintains a constant orientation with respect to its source and the fiber material through which it passes. Noncoherent light is random in dispersion.

It is said that fiberoptics systems can provide very precise control of a robot's arm movements, hand movements, etc., since they measure so precisely the position or movement of these extremities. Also, the light rays used with optical fibers provide the simple light-dark or one-zero logic which is easily adapted to computerized control. Of course, you can imagine an interrupted light beam as having meaning also (recall the photoelectric and infrared announcing systems such as sold by Radio Shack and other electronics suppliers.)

Because optical fibers are so small, you might use many of them to give some idea as to hand pressure exerted by a robot or the shape of objects as "seen" from the hand position. (See the TAB books on robotics (numbers 1071 and 1421) for a more complete discussion of robots and this phenomena.) Fiberoptical switches are easy to imagine. An off position occurs when you move the light-carrying fiber a little away from another so that the light does not go through the second fiber. Moving the first fiber back so it sends light into the second fiber then makes the light path continuous to the detector and this becomes (or can become) an on position. Of course the off and on positions can be changed by reversing the wiring.

Because they are flexible, optical fibers can be used with moving parts of the robot without problems. Two such optical fibers can be used to guide a machine tooling device along some kind of pattern or edge of a work-piece. They offer the robot long-range vision as the actual pixel element formation unit doesn't have to be at the work end of the fiber which is scanning the illuminated object.

One of the advantages of using optical fibers to transmit light rays to dark areas and thus illuminate them (as in arthroscopic surgery) is that cold light can be used with high intensity. If a large area is to be illuminated, many optical fibers are used, and placed and adjusted so that shadows and dark areas would not exist. Heat is generated when the conversion device (the light bulb) located near the work object does not convert all of the energy into light rays. Some of that energy is converted into heat. When you can have the source of the light a long distance away from the object, the light rays as conducted by the optical fibers do not have that heat content (unless they are near the infrared realm where heat is automatically produced with the ray); this is called cold light. Fluorescent light tubes produce cold light, but there is a lot of heat generated in the transformer which supplies the high voltage that is used with such devices.

Since the small size of the receiver and transmitter unit of a light system permits close proximity in confined spaces, the various applications of light are multiplied enormously. Figure 4-10 shows how such units can be combined in a single block to measure some size, shape, or movement of a machine such as a robot. Figure 4-10 shows that the robotic hand has optical fiber nerves that connect to the LEDs and the photodiodes (receiving units). These nerves connect to various amplifiers, computing elements, and integrated circuits that can analyze and determine what the robot's fingers see, how they are grasping something, where that something is with respect to the robot's hand, and so on. Actually, each LED shown might be a dual light source which then gives a differential output for each finger. The possibilities for hand instrumentation using fiberoptics and associated devices is almost unlimited.

Many professionals feel that the use of LEDs, lasers, and infrared beams of light conveyed by optical fibers may be the answer to some perplexing

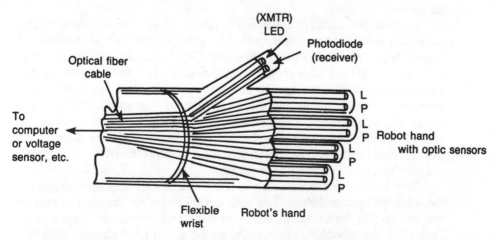

Fig. 4-10. A block of transmitter/receiver light elements in an application in a robot's hand.

problems in connection with the transmission of light. Infrared beams are not subject to interference by ambient light rays. They work equally well in daylight and darkness. They are invisible to the human eye. Thus, if the light changes in an environment wherein the sensor is used, that makes no difference in the operation of the machine. Some believe that small amounts of dust may not affect the transmission of the infrared beams. If the part being handled is hot enough to emit infrared radiations than it is easy to follow and track its position with infrared sensitive sensors.

THE DISPERSION PROBLEM WITH OPTICAL FIBERS

There are two possible modes of transmission of light in an optical fiber. One is the single mode, the second is the multimode. One equation that indicates how many modes might be propagated in a given optical fiber can be stated as

$$f = \left[\frac{\pi d}{\lambda} \sqrt{(n_1)^2 - (n_2)^2} \right]^2 / 2$$

where n_1 = index of refraction core, n_2 = index of refraction cladding, d = core diameter (meters), λ = wavelength (meters), f = modes propagated.

If only a single mode is transmitted, then this equation is not applicable. Single mode propagation occurs with a very small core (only a fraction of the size of a human hair). This makes joining a number of fibers together end-to-end (to make up a long run of such fibers) quite a formidable task. The dispersion problem, simply stated, is that various frequencies, and thus, wavelengths are propagated through the fibers when conditions are not perfect. As you know, things can never be perfect. The reason many modes might be transmitted is because the source of the light (say a LED) may actually create several or many frequencies within the very narrow band found at these tremendously high ranges. If the guide is so designed that it can sustain these other frequencies, and they are strong enough, then they propagate through the optical fiber guide.

Mode dispersion problems can affect the transmission of any intelligence transmitted by means of light pulses. This is sometimes referred to as the impulse response capability of the optical fibers. Mode dispersion and material dispersion tend to broaden the light pulses time-wise and even though you might have begun the transmission of intelligence with short and well spaced pulses, this type of dispersion can cause the pulses to occupy a long time interval and thus reduce the time space between them until in the worst case the pulses actually overlap and there are no pulses and spaces existent at all—There is just a solid continuous ray of light at the exit. Dispersion then relates to the speed of propagation of the various frequencies within the transmitted band of frequencies in the optical light guide.

Although much literature talks about modes, the precise definition of a mode is lacking. A mathematical equation might be used to describe a single-mode or a "multi-mode" optical fiber system; after analyzing such an equation, you can conclude that one or several frequencies or wavelengths seem to be propagated depending on whether the system is single-mode or multi-mode. A dictionary definition of mode is: "A particular form or variety of something." It might also mean "style." Mode and dispersion are important terms in fiberoptics so let us have another go at understanding them.

Mode is related to the number and variety of wavelengths that might be propagated through the core of an optical fiber. The mode is also related to the velocity of propagation of these wavelengths that might be propagated through the core of an optical fiber. It is also related to the velocity of propagation of these wavelengths through that core, normalized, as they say, or related to a one kilometer length of fiberoptic material.

If you know that a LED used to excite a beam of light in an optical fiber actually sends a band of light frequencies into that fiber, then you will have no trouble grasping the idea that there are various wavelengths or wave fronts moving inside that transparent material. You should also have no problem considering the fact that some of these waves may travel in a direct line through the fiber and some will bounce around by being reflected from the junction of the core and cladding. These require a longer time to get from the input end of the fiber to the output end of the fiber. A LED may have a band of plus or minus 25 nm about a center frequency. It generates this band just because of the way it works. There are a number of wavelengths trying to go through the optical fiber when it is excited by a source device. Some do not get through because they enter at such an angle that the critical angle for propagation is exceeded and hence they escape into the cladding. But lots of others do make it into the fiber, entering inside the required acceptance cone angle, and propagate in one way or another along the fiber length.

Modes are discrete incidence angles for the light rays. A glance at Fig. 4-11 explains this further. Figure 4-11 shows a direct ray going straight ahead from input to output (a). Then there are some captured rays which hit the junction between the core and the cladding walls at less than the critical angle for escape and so are reflected (b). Some rays are too direct and, having passed the critical capture angle, escape into the cladding and are lost forever (c). It isn't hard to look at this sketch and say, "If there was just one light frequency (such as from a laser) then it might be the direct ray, and since it passes down the core, it is single mode." But perhaps this is optimistic. Is there anything which produces just one frequency with no harmonics? The initial ray from the course (c) has been drawn as a direct line from the source. Actually, since there is an index of refraction between the air and the core, there is a one-bend kink in the first part of the ray (c) which is not shown.

Since the higher order modes travel farther (due to reflection or bouncing around) they must arrive at the output after the lower order modes arrive. In this context a higher order mode must be a lower frequency since it has a longer wavelength and thus bounces around more — or could it be that it has a higher frequency which produces a shorter wavelength and this is refracted more? Think about it.

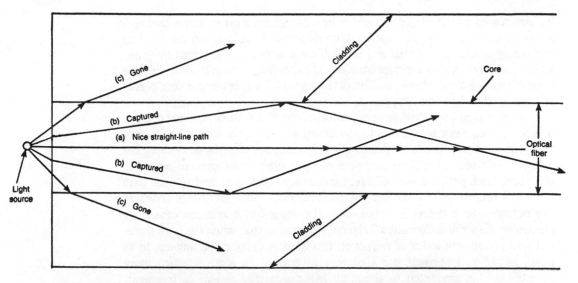

Fig. 4-11. The transmission modes shown are various incidence angles of propagation.

In any event, for nice sharply-shaped pulses require many frequencies all put together. In this context, if you say that some frequencies from a pulsing light source will not get to the fiber output at the same time that other frequencies do, then there are some delays in transmission. That has to affect the pulse shape at the output (make it rounded, less well-defined and so on). This is no problem if you are using a slow transmission of pulses, but if you are in the 50 megabit range or higher then you've got a problem. Delays such as this cause errors and loss of pulses and pulse shapes.

These delays have a name: chromatic dispersion. Simply stated this means variable delays due to differing wavelengths trying to propagate through the optical fiber. It has been found (by Northern Telecom) that pulses of light in the 840 micrometer region are broadened about 0.1 nanosecond (10^{-9}) per kilometer of fiber per nanometer of source width. Thus a 5 nanosecond pulse from a LED which has a 40 micrometer spectral width will, after going through some 5 kilometers of optical fiber, broaden out to about 20 nanoseconds. A laser source pulse which has a 2 to 4 micrometer width gains a broadening effect of about $\frac{1}{2}$ microseconds. This reflects at least one problem in the transmission of high speed pulses with light. These delays caused by the fiber material and the different transmission times of the various independent frequencies through the fiber, result in changes such as broadening of the pulses and compromising the pulse shapes in this method of transmission.

There is another effect called multimode dispersion. This is a concept which you can grasp quite easily if you remember that there is such a thing as interference between one wave and another in the electromagnetic spectrum. Remember, radio waves are reflected from the Heaviside layer (if the angle of incidence is less than the critical angle) and they are reflected and fall upon various places on our planet and add to or subtract from other waves that also arrive at that particular destination. With an add-to situation, you can have an increasing intensity of signal and in the listening device the sound becomes louder (usually). With a subtract-from condition, fading occurs; if conditions are exactly right, a complete elimination of the signal for a short or medium period of time at that location happens.

In an optical fiber, that same condition exists, but at the frequency of light. The same frequency waves tend to add to and subtract from each other as they progress down the fiber's length. Only a given number of paths or modes exist in the fiber which permit transmission or completely from input to output of the fiber. Each path for each different frequency is different and so each path (direct or reflecting from the walls of the core) has a different time of arrival at the output—or a delay. Then we have the same bad conditions caused by chromatic dispersion. Northern Telecom discovered that when the core material was of uniform index of refraction this kind of delay could amount to as much as 20 nanoseconds per kilometer of travel. To state it much more scientifically, it amounted to about 20 MHz/kilometer roll-off of frequency meaning about a 3 dB optical bandwidth. This in turn meant the loss of those wider bands of frequencies necessary to make up a precise pulse shape. A square-wave pulse mathematically requires an infinite number of frequencies, harmonies, etc. to be a square-wave pulse. That says, in turn, that we need an infinite bandwidth. So, if the bandwidth is reduced, the pulse shape is affected. In radar, the bandwidth required is

$$1/\text{pulse width}$$

and this little expression gives us some idea of what is going on in the narrowing effect of the multimode dispersion in optical fibers. There is a word used in connection with this effect: intramodal. If a graduated system of "index of refraction" is fashioned within the core material, this results in increased bandwidths as high as 3 to 6 GHz/km with some experimental types of fibers.

Some good effects also occur in the transmission of light rays through an optical fiber core. Mode mixing which can take place across splices and within the fibers themselves. When the paths of the various light frequencies are different, it takes different times to get different rays from the input to the output. In mode mixing there is some interplay between the various modes (paths) in such a manner that the time it takes all rays to get from input to output tends toward an average value. That means that the delays tend to balance out and all rays tend to arrive at the output at the same time. How nice. Well, to be honest, you cannot have everything, and so there will still be some delays, but these won't be as bad as they might be if this mixing phenomena didn't exist. It is convenient at times to induce mode-mixing by messing up the light-fiber path. Using microbend effects, which result when you twist and bend the light fibers, really confuses the rays so they don't know which path they had a moment ago and use new paths so that mixing occurs. The problem is that increasing the path length this way and mixing the modes causes some losses. Well, it is bound to happen, when you confuse the path and some rays don't know where they are going, they go on an unpredictable manner and get lost. It is a trade-off. You have to consider the good effects along with the attenuation and the sources of light and then decide if you want to improve this effect or not. Perhaps more study and experimentation will determine the proper way to go. Remember that intramodal is a term meaning inside a fiber mode or along a fiber ray path, while intermodal means the various paths from one fiber to another.

In most college physics courses much time is devoted to light and the use of diffraction gratings and lenses and the concepts and mathematics and experiments concerned with optics which are appropriate to our discussion. One such series of experiments involves a prism and diffraction grating (Young's experiment) to determine the lines of the optical spectrum. Each line is as a band of light frequencies, so closely related that they seem to merge together to become one central frequency or line. It is common to describe the light rays transmitted through an optical fiber as being in "lines," meaning the wavelengths which actually pass from input to output through this type of medium. (For more information, see any good, standard college textbook. I highly recommend a serious study of Young's experiments if you are looking forward to a serious study of optical fibers. Michelson's interferometer also is recommended as a basic study in that our modern laser gyroscopes are based in principle on this experiment.)

Let's get back to the use of a diffraction grating and/or a prism. When an incident light is focused on such devices, they spread out the light spectrum so that one can see the various bands, or lines of light. The prism produces colors which identify with various frequencies. The diffraction grating produces light and dark lines showing the reinforcement of rays or cancellation of rays as a function of distance. The resulting patterns can be analyzed to obtain various information about the light ray being propagated. Diffraction gratings are often used to measure wavelengths and to study the structure and intensity of spectrum lines. This can be important in the study and experimentation efforts concerned with optical fibers.

With optical fibers, even of a transparent plastic, light can be piped from one location to another by introducing the light into the fiber at one end in such a manner that total reflection will occur at the output boundary of the rod (where the cladding or strengthening material joins the rod surface) if and only if the light is introduced at an angle below that of critical. The critical angle, that angle at which the light hits the boundary edge of the two materials, is less than that which permits the light to pass through that boundary. Mathematically that critical angle in terms of the material is:

$$\text{Sine } \Theta_2 = \frac{n_2}{n_1}$$

where n_2 = second medium index of refraction for air equals 1.00 and n_1 = first medium index of refraction for glass of a given type is 1.50.

Thus the critical angle is that angle whose sin is 1.00/1.50 or 0.667 and the angle is 41.7 degrees. This means that if the light rays hit the boundary or edge of the glass rod at an angle, measured from the perpendicular across the rod, at less than this angle, the light escapes through the rod's edge into the second medium. (See Fig. 4-12.)

Unless the incident angle is 90° (at which angle all the light rays pass through the rod into the second medium and escape and do not travel down the rod) some light rays are reflected down the rod even if the angle is less than the critical angle. The closer the incident angle gets to 90° the fewer the number of rays which are reflected, until at 90° none go down the rod length.

Since images are variations in light intensity which impinge on our eyes, if you can make these variations in light come back through an optical fiber or a bundle of optical fibers then you can see things which the other end of the fiber is pointed at, providing that the objects there are properly illuminated. This is the basis for the medical instruments which permit examination of the human body's interior. One bundle of fibers can be used with magnifying lenses to see what is going on inside.

When optical fiber bundles transmit images, each fiber transmits only a small segment of the image. It is common to use these bundles of fibers (in which each one carries only a small part of the image) to convey a scene from one end of the fiber length to the other end. The ends of the fibers on the receiving end are all viewed simultaneously so the image can be seen. This is like building up a picture on your TV screen by using a multitude of dots. You aren't aware of the dots; When you view the screen we simply see the total image.

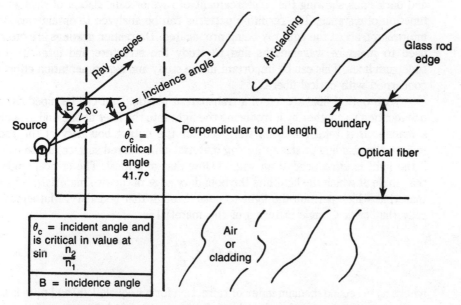

Fig. 4-12. The critical angle of reflection inside a fiber-optic waveguide.

Since the fibers may be flexible, and they are very small in diameter, they may be clad in some strengthening material so stresses and strains won't affect them. If you've visited any gift shops in recent years, you have probably seen a display of an optical-fiber-tree. This unit, rather pretty in concept, with its flowing lines of almost clear transparent material emerging from a vase or jar or whatever, has at the tip of each fiber a glowing dot of light. That is the reflection of the light source inside the container, and it appears only at the end of the fiber. Inspecting such an ornament may give you some ideas as to why so many fibers are necessary in a bundle to convey a total image from one end to the other. If you have a chance, examine the fibers as to size, material, light emission and so on. You'll find that there is a light loss due to the absorption of light rays by the material itself.

The image transmission problem has produced several solutions. If some fibers are tapered so that they have a large-end and taper down to a smaller end, the larger end can be placed to view the image. As the light rays come back through the fiber the image is carried intact and can be seen with relative ease using magnifying lenses that may or may not be attached to the viewing end of the fibers physically. Another phenomena occurs in the transmission refraction and reflection of light rays called double refraction. It is important because it can mean that a single ray may be split and have two sections or beams that are polarized at right angles to one another.

Study any good college physics text for details of this phenomena. The fact that light can be polarized can be of the utmost importance. Light tubes can be designed that accept only certain polarizations and not others, giving light selectivity. Also, polarizing two beams from the same source so that they have equal and opposite polarizations, then cancellation can take place when they are combined. This effect is useful in a light communications system. A calcite crystal can polarize two emerging rays from a single ray and so they have a 90° polarization with respect to one another. An optical fiber system must not cause cancellation of the rays within itself or the losses would be too great to make such a transmission system practical. The next chapter examines some communications applications using optical fibers.

Chapter 5

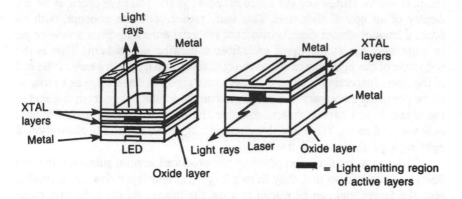

Light rays

Metal

XTAL layers

Metal

XTAL layers

Metal

LED

Light rays

Laser

Oxide layer

Oxide layer

■ = Light emitting region of active layers

The Use of Optical Fibers in Industrial Applications

ONE OF THE MOST IMPORTANT CONSIDERATIONS IN THE USE OF OPTICAL FIBERS and light rays for communication and data transmission is the fact that light rays are almost immune to electrical interference when sent over an optical fiber transmission path. Electromagnetic radiations such as sparks, lightning, and crosstalk are practically eliminated as interference sources in an optical fiber transmission system. Think how important this is! A large computer installation, for example, requires communication among and from the various machines, and the integrity of the communication is dependent upon elimination of interference signals from the system. Using light and optical fibers as the medium solves this problem.

Consider the meaning of the use of light frequencies in communications. A small band of frequencies (kilohertz width perhaps) is needed for transmission of intelligence; just think of how many such bands can be contained in the light region of the frequency spectrum without interference from each other. Also, since the bands might be made wider, information can be transmitted at much faster rates. Gigahertz rates and perhaps even higher rates might be used and still there are sufficient bandwidths to handle a very large number of simultaneous channels. Optical fiber systems can send analog and digital data side-by-side without causing any problems. What this means to telephone companies is evident. Costs are less than with copper wires. Reduced crosstalk, interference, and fewer cables mean that it is almost inevitable that sooner or later all telephone communications channels will use this means of transmitting data, telephone, telegraph, and video signals.

THE THREE BASIC ELEMENTS IN A FIBEROPTIC SYSTEM

There are three primary elements required in a fiberoptic system for communications. The first element is the transmitter. That is the unit which must generate the light rays and be capable of being switched on and off quickly and/or modulated with some signals representing intelligence. The second is the optical fiber which must have purity, cladding, and packaging so that it is strong and transparent to the light frequencies being used. It must be capable of being spliced and repaired when necessary (at a reasonable cost) and be able to convey the light rays a reasonable distance before a repeater station has to reamplify the light beams to enable them to traverse the total distance over which they must travel. (In some cases, many repeater stations might have to be used.) Third, the receiver element must reconvert those light rays back into analog or digital currents and voltages so that the user station can separate and use the intelligence signals being transmitted.

In some cases control signals of a dc or low frequency ac voltage have to be used in conjunction with the light ray information. Such signals may have to be sent over copper-type lines. Light rays cannot transmit dc signals as such. However, it is entirely possible to send digital data in the form of very high frequency pulses over the light beams and then reconvert these into appropriate on and off signals using a decoding device at the receiving end. Slight variations in pulse coding, width, spacing, and so on might be used with light beams to establish control functions and directional signals for control or machines or devices at a remote location. There are now on the market transistor integrated circuits which convert analog-to-digital and digital-to-analog signals. So, once you reconvert the light rays or modulated light ray signals into voltages and currents, the solid-state technology is here and available to do whatever is desired with the signals in this latter form.

OPTICAL FIBERS AS SENSING DEVICES

If the optical fiber is stretched, bent (even microbends), subjected to temperature influences, or even subjected to the influence of electromagnetic waves which can change the polarization of the light rays in a manner proportional to the magnitude of the disturbance, there are sensors of those disturbances. If the effect can be mathematically explained, a quantitative output of light can be produced that is proportional to the magnitude and/or direction of the effect. For example, suppose you place an optical fiber near a heat source. As the temperature increases, the fiber expands diametrically which changes the path length of the rays going through it and the angle of reflection and refraction. This causes a change in the amount of light (or polarization of light) passing through the fiber. You can measure the light intensity at the receiving end, and relate this in some proportional manner to the temperature impinging upon the fiber. Voila, a heat sensor.

Similarly you can split the beam of light passing through an optical fiber. This light beam is produced by a laser so that the rays are uniformly directed and with a given polarity (E vector). One half of the beam is used as a reference or standard to which no changes occur. The other half is sent through some device influenced by the outside world phenomena such as sound, light, pressures, temperatures, motion, and so on. By comparing the change in polarization of the affected half of the beam to the reference beam, you can measure the difference in polarization change and relate this mathematically to the world phenomena changes.

Optical fibers can and are being used as sensors in a variety of ways, object positioning, determining object size and shape, determining rotation direction and rotation rates, measuring temperature changes, monitoring and detecting sound by either an interferometer or pressure-change method, and even to detect electric currents and/or voltages. These latter sensors use the change in polarization of a split beam caused by the electromagnetic field of an unknown current and then comparison with the standard as just described to determine the magnitude of the current.

SOME DEVICES AND EFFECTS USED WITH OPTICAL FIBERS

There is an effect associated with light rays or beams transmitted through optical fibers that you should be familiar with called birefringence. What the word means is that the fiber transmits a given polarization of light rays of a given frequency faster than another polarization of those same light rays. Thus, when some effect happens to a split light beam which causes the polarization change in the variable half of the beam, a phase shift occurs upon recombining the two beams. This phase shift can be accurately measured.

Electromagnetism is measurable through this effect. A magnetostrictive core material, with an optical fiber wound around it, expands and contracts or vibrates when exposed to some magnetic field of varying intensity. The optical fiber changes shape slightly. This causes some parts of the light beam passing through it to have a slightly different polarization than the rest of the beam. Light rays are electromagnetic in nature and so the fact that they can be influenced by other magnetic and electric fields is not surprising. By passing the beam through the proper Wallaston prism, the parts of the beam with a given polarization can be separated and then compared. This is a means of sensing the magnetic field intensity, frequency, and so on.

POLARIZATION

Now that the subject of polarization of light rays has been introduced, let's examine this subject in slightly more detail. Light is an electromagnetic radiation. It is a transverse wave that has its electric and magnetic vectors at right angles to each other and to the direction of the wave's propagation. Normally such a wave is described by the poynting vector shown in Fig. 5-1. The diagram shows the relationships between the electric field, the magnetic field, and the direction of propagation of the wave front. Often, the symbol $\hat{B}$ is used instead of $\hat{H}$ for the magnetic component of the wave.

The direction of the $\hat{E}$ vector at any time (t) tells us the polarization of the wave. When the $\hat{E}$ vectors of a wave align in a given direction, or plane, that

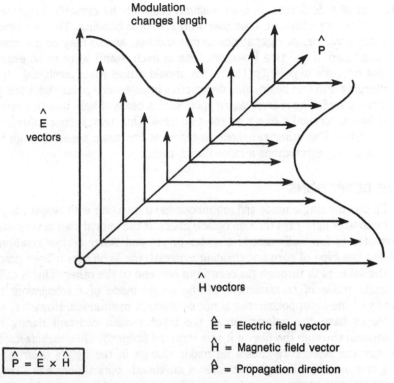

Modulation
changes length

$\hat{P}$

$\hat{E}$
vectors

$\hat{H}$ vectors

$\hat{E}$ = Electric field vector

$\hat{H}$ = Magnetic field vector

$\hat{P}$ = Propagation direction

$$\hat{P} = \hat{E} \times \hat{H}$$

Fig. 5-1. The poynting vector of electromagnetic radiation with modulation of the (E) vector.

plane is said to be the plane of propagation of the wave. In a plane-polarized wave, all such planes which contain the $\hat{E}$ vectors of the light beam are parallel. The electric and magnetic vectors of Fig. 5–1 are extended at right angles to the direction of propagation of the wave. Even when these vectors vary in magnitude, as they do with modulation of any type or kind, they still have this orientation to the $\hat{P}$ vector at any and all instants in time.

Under some circumstances, there may be a circular polarization condition that exists if the plane of polarization of the wave is caused to rotate around the $\hat{P}$ vector. There can also be conditions where a single light beam may be split into two sections, each of which may then have a different polarization due to some external causes (such as the sensing of some external and changing phenomena as described earlier). You need to be able to determine the polarization and change of polarization of light rays in some manner.

In 1817 a scientist named Thomas Young was investigating the properties of light and of sound (which produces a longitudinal, or pressure type wave) and performed an experiment by which light rays could be investigated. He found that light rays had the transverse condition of propagation just described. Two of his contemporaries, Arago and Fresnel, were able to split a light beam into two sections or paths by allowing it to pass through a calcite crystal. This production of a double beam from a single beam at the crystal's interface is called double refraction, or in simpler words, double bending. The two beams have their Ê vectors at right angles to one another, and so they do not cancel when combined. if the two Ê vectors (one in each beam) were to be exactly equal but opposite in polarity (180°) they should cancel when combined. Upon their merging into one beam, total destructive interference occurs with the net output being null. The reason calcite splits such a beam of light into two rays is caused by the crystal's effect on the two rays. One travels more slowly or refracts more. The other ray tends to travel at the same speed through the crystal and hence traverses a more direct path.

SOME DEFINITIONS

The terms single mode and multimode have meaning with respect to the transmission of light rays through optical fibers. If the optical fiber is very small in diameter (a few millionths of a meter or so) and under certain conditions involving the type of core and cladding material, the light rays follows practically the same path through the core from one end to the other. This is called the single mode of transmission. In the single mode of transmission, the integrity of the input polarization is not necessarily maintained. However, it is possible to have the polarization of the input remain constant during the transmission through the fiber if it was stressed properly when manufactured. This way the fiber core causes no major change in the light's polarization during transmission through that core. A multimode optical fiber has a larger core and the light rays travel many different paths from input to output depending on their frequency, wavelength, and insertion angles. In much of the literature there are references to step index types of fibers and also to graded index types of fibers. A step index means and abrupt change in the refractive index of core and cladding. A graded index means a gradual change in the core index of refraction that is accomplished by changing the core material in some graduated manner from the center outward to the cladding boundary.

DETECTING SMALL CHANGES IN THE PHASE OF LIGHT RAYS

There are many industrial and military applications of optical fiber systems requiring very small changes in phase existing between two light beams of the split-beam type to be detectable and measurable. This is no small task. There can be phase changes due to reflection, or there can be phase changes due to the change in the fiber structure through which the rays pass. These latter changes can be due to such phenomena as water pressure in advance acoustic systems and magnetostrictive effects on substances that have physical contact with the optical fibers (heat and so on). There can be phase changes due to actual physical motion of the base structure of the optical fiber system as is found in the laser-type gyroscopes.

To determine these very small phase changes one uses an interferometer —and that brings to mind the famous Michelson-Morley experiments discussed in most good physics texts. With light, use a two beam system to

116

compare the phase of a variable-phase beam with a reference beam. In this system of comparison the light beams must be coherent or have the same polarization; this is obtainable from a laser. However, some studies point out that the coherence of the beams may change depending on the reflections (paths) internal to the optical fibers for the frequency being propagated.

Basically, the interferometer effect is a reinforcement or cancellation of the wave fronts. Therefore, you could propagate almost any type of wave, split it, and cause some shift of the phase of one beam due to something external or something to be measured. When recombined, you can see how much interference one beam has with the other under these conditions. By some analysis it is possible to relate the interference effect with numbers that represent the changes to the external world being measured so that you can quantitatively measure those changes.

The optical fiber system of light transmission is flexible to some extent. Imagine a splice between two sections of optical fibers. When the two ends are perfectly aligned, the maximum amount of light is passed from one section to another. If anything causes any shift in this alignment, then less than maximum light transmission takes place. Aha! Now this is a way to measure anything from physical vibrations to detection of objects in a robot's gripper! If one section of optical fiber is permitted to vibrate with the outside force, the changes in light are proportional in frequency and amplitude to that vibration regardless of its cause or source. The sudden darkening of light transmission from such sensors in a robot's gripper could indicate to a computing section that the gripper has located a part, or encompassed a part to be moved. The requirement here is that the light through one section of the optical fiber must be constant and the second section of the fiber must be permitted to vibrate with an external force through a small displacement. The system might have to be capable of transmitting light rays through an air path as would be necessary going from a robot's fingertip to its thumb so that interruption of the beam of light can mean that an object is intervening between those two robotic elements. The reflection of light rays (as in a grocery store bar counter) also has meaning in this context.

Almost anything can be sensed using the phenomena of polarized light shift, or intensity shifting of the light rays, when the proper conversion element can be fabricated. One example of this is the Pockels effect, which is simply the rotation of the light beam's polarization as it passes through a crystal subjected to the pressure of an electric field due to an applied voltage across that crystal. You know how a piezo-electric crystal squirms under an applied voltage. It is opaque to light so it won't work in this application. However, there are some crystals of other substances which are clear to light transmission and cause some effect on the light rays due to the application of voltage to the crystals. Lithium Niobate crystals are one way to fill the need for this type of sensing device. They cause a rotation of the polarization of the light ray that is proportional to the magnitude of the applied electrical voltage. Now, all that remains is to use some kind of intereference-effect detector system to compare this rotation amount to a standard that has not been rotated. By using a mathematical relationship between the rotation of the polarization in degrees and the applied voltage, you can determine the voltage by measuring the rotation. The interferometer offers a means to determine the polarization rotation.

LIGHT AND COLORS (MONOCHROMATIC LIGHT)

You can visit your local Radio Shack store or equivalent purveyor of electronic devices to examine the ICs, LEDs and optoisolator elements that have to do with light. LEDs are available in various color outputs such as yellow, red, and blue. These LEDs are monochromatic (one color). It also means one wavelength, but let's not worry about that here since we identify a color with a wavelength in the visible light spectrum. For identification purposes, if a light is not monochromatic, assume that the light has many colors. White light, for example, is composed of many colors.

In television, there are three primary colors: red, blue, and green. The manner in which these are mixed gives rise to a multitude of other colors, sufficient for color television. If you have ever adjusted your TV color controls, you know that the proper mixing of these colors gives you a grade of white ranging from grey to white white. With one color, you have a monochromaticity and when you combine colors we have nonmonochromaticity.

Monochromaticity is important in the world of optical-fiber light-transmission, containment, handling, and the use of light in control, measurement, and sensing applications. If light is monochromatic then the optical fiber can be designed so that it transmits that particular light best, and it doesn't matter if other wavelengths are not transmitted or handled as well. In the visible light spectrum it is nice to glance at the end of a fiber (obliquely and never directly at the end for safety reasons) and seeing the color you can say "Well the wavelength of that light is about . . ." and you fill in the appropriate micrometer dimensions.

The U.S. Navy is very interested in optical fibers and their use and began a program in 1978 called FOSS, an acronym for Fiber Optics Sensor Systems. It is a big research and development program and most of it is classified. But knowing that this arm of our Department of Defense is vitally concerned with this material tells us how important fiberoptics will be in the future.

LASER WEAPONRY

The military establishments of the world are constantly trying to develop a laser of high enough power, small enough size, and economically feasible enough to use as a weapon or in a weapons system. Already laser light, PIN, and avalanche-diode detectors have been used in precise ranging systems. These determine the distance to an object by pulsing the laser light at it and timing the trip there and back, like radar does when it determines range. There have been reports which imply that laser weapons in outer space have been used to disrupt and destroy satellites in tests—highly secret of course. In space there is little or no attenuation of the laser beam signal.

Only if that beam can be focused precisely (as for retina operations) is the tremendous energy contained in a laser available for death and destructive purposes. Otherwise, the light is there and in its modified and weaker existence can be used in many other applications such as transmission of data, communications, pictures, and so on, over fiberoptic links and in some ways in military applications.

John J. Fialka, writing in the Wall Street Journal, has stated that in mock battlefield games: "The weapons on both sides will be equipped with lasers; the firing of a blank round will send an intense, pencil-thin beam of infrared light at the enemy. The soldiers and vehicles will be equipped with button-sized electronic sensors that react when struck by the laser beam. A "kill" on a soldier will cause a steady beeping in a little box he carries; the sound of which can only be turned off by a key attached to the soldier's rifle. When he uses the key the rifle no longer works. The soldier is out of action until the umpire sends a special electronic impulse that unjams the system. Some soldiers and tanks will be equipped with special small black boxes which contain transponders (special devices which can be queried by a specially coded signal). These black boxes will record when weapons are fired and what kinds of hits are received. They will also send out position signals to some number of antennas surrounding the mock battlefield. A computer system will interrogate the black boxes 10 times per second, analyze the data, and give base commanders and technicians a running account of the battle." These laser weapons would be practice weapons to improve a soldier's ability to register hits on an enemy target. Such a system has never been possible. Little did Mr. Fialka realize that he would be describing the "toys" currently being hyped by manufacturers. Laser weapons with shields/detectors are quite the fashionable game of today's youth.

In a laboratory at the Johnson Space Center, one scientist was concerned with running an electric toy train back and forth on about 300 feet of track. As the train moved, so did the bicycle reflector mounted on the engine. This reflected the pencil-thin beam of a surveyor's laser instrument. The information from the reflected pulses was used by a small computer to present a digital readout of the distance, at any time (t), to the engine. The purpose of all this experimentation was to develop a docking system—automatic and self guided —for use with satellites and other spacecraft. The laser used was a gallium-aluminum-arsenide semiconductor about the size of a grain of salt. This laser emits light in the infrared region just past that region that is visible to the human eye. You cannot see the beam but it exists. Five tones modulate the light beam and are used as the basis for a comparison circuit to determine time of travel and position (with accuracy within millionths of a meter) in determining the target's location and identification. The experimenter indicated that this system required some auto-reflecting material to bounce the laser beam back to its source for the system to function properly.

Radio Shack makes an infrared pulsed-light announcing and security system that in many ways duplicates this concept. It provides the sources of the light (laser) and a receiver in close proximity, and a special grating which can be positioned to reflect the infrared beam. The nice thing about this system is that ambient light from the sun or incandescent lights does not affect it, and you cannot see the beam — but don't look into the laser lens to try! Instead of a computer which provides distance information, the system sounds an alarm if the beam is broken or interrupted.

SOME COMPARISONS BETWEEN COPPER
WIRE CHANNELS AND FIBEROPTIC CHANNELS

Bell Northern Research Limited of Canada has done some study and made some comparisons among some methods of handling long-range telephone communications. These results are of interest when considering industrial applications of the fiberoptic communication and data handling systems. When considering radio communications systems, the cost rose very high due to the necessity of having access points at about 16 kilometer distances along the routes. Also, the system had to be bidirectional for two-way conversations. Analog information systems picked up noise and garbled the speech, and digital techniques were ruled out because of the non-availability of the coaxial cable carrier systems needed for this technique. With either analog or digital systems a large number of high-cost repeater stations would be needed, and susceptibility to ac interference would be very high.

In typical copper carrier systems up to 1344 voice systems or channels are digitally routed by copper lines to radio sites where they are multiplexed so they can be handled by radio carrier systems. When considering the use of fiberoptic channeling and carriers and using a multiplexer at the nearby toll office, it was estimated that only four fibers (plus spares) would be necessary to handle all the channels of voice communication previously carried by up to 115 copper wire cable pairs. This resulted in a considerable cost difference. Recall that the optical fiber being considered here might be only 125 microns in diameter — and that is small indeed!

HANDLING LONG RUNS WITH OPTICAL FIBERS

Over long distances losses and cable, fiber, and wire breakage must be considered. The losses due to the transmission system are compensated for by repeaters and/or amplifiers of a suitable type. The breakages involve splicing and rejoining the ends of the wire or fiber. When reamplifying the signals, repeaters must be located at reasonable distances to prevent the losses from getting so large that the signal intelligence vanishes and can't be reamplified. You have to be able to detect the signals clearly to amplify them and send them on their way.

The splicing of cables or fibers may present a different kind of problem. Splicing wires is easy. You just find the break, twist or mechanically fasten the ends of the broken line together, solder them for increased electrical conductivity, and then reinsulate the joint. Mending a radio channel may involve replacing a circuit board in a transceiver, adjusting the antenna, improving electrical connections in the system, or repairing the multiplexing computer — a somewhat more difficult job.

Repairing a fiberoptic line involves making a perpendicular end on each of the two sections, then bringing these together for a perfect pressure fit and keep the two segments so connected for the balance of the time the system fiber is used. Considering the fibers are 300 microns in diameter (with cladding) and the parallelly-cut ends must fit perfectly and tightly, this is quite a problem. Bell-Northern solved this problem by inventing a machine to make that splice automatically.

An optical fiber cable may have just optical fibers or it may also include some copper wire conductors for tests and measurements. The cables themselves may be subjected to various chemicals which act upon the cable material in a nondesired manner. There may also be some effect from nuclear radiations, but this isn't likely. All cables are subjected to tension, twisting, crushing, and vibration.

The cable must be designed for a minimum of loss. Bell-Northern has achieved less than 3 dB/km attenuation at 840 nm wavelength (nm is nanometer). This is a cable that has a flexible core of steel strands covered with a plastic that is grooved to accommodate the optical fibers and copper wire pairs. There are separate grooves for control, measurement, and test purposes. A dc current is usually used for control purposes but light rays have to be modulated and demodulated to produce the same selective controls. The cable must be wrapped with a material that can withstand the environment whatever it may happen to be (a stretch of wilderness with extremes of cold and heat, a factory with chemical pollutants, a stretch of underground city street, with the changes of pressures and stress and strains and so on).

Once you have sufficient knowledge of a subject to understand it from the theoretical viewpoint, you may want a little hands-on experience to go along with that theory. If you now have that feeling, then look at Fig. 5-2. This shows a very simple system (using parts readily available from Radio Shack) with which you can do some experimentation with light communications and even light control systems according to your ability to invent, imagine, or design these circuits.

The LED and receiver diode are from Texas Instruments, but equivalent types can be obtained from Radio Shack. You will have to do some experimentation with the circuit shown. The plastic optical fiber is obtainable from Edmund Scientific Co. They have various cables that you can get to experiment with. For example, you might use a multi-cable of several strands of optical fibers, and then use various colors of LEDs to see how the different wavelengths propagate through the fibers. If you then figure out a way to pulse the LEDs with some form of code, you might design a control system so it can close some kind of relay (solid state or mechanical). It's a fun project for the dedicated experimenter.

Along with the propagation of light rays through optical fibers, you might want to experiment further with a system using lenses and fibers. Edmund Scientific has such a device called a fiberscope. It has coherent optical fibers with a 7 magnifying eyepiece. This eyepiece focuses from infinity to as close as $\frac{1}{2}$ inch.

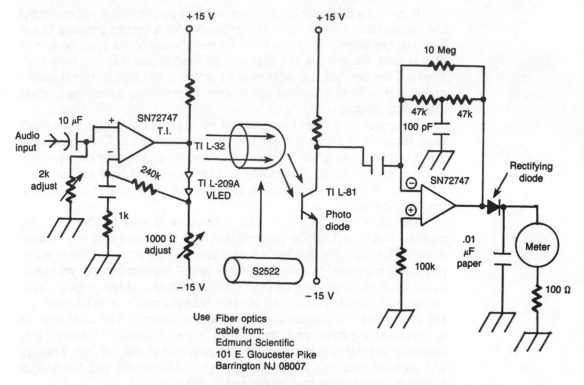

Fig. 5-2. A simple communications system using light beams.

USING LENSES WITH OPTICAL FIBERS

You may have already considered using lenses and wondered why it was not mentioned sooner. It seems logical to use a focusing lens to get light into an optical fiber guide with as little loss as possible. Almost everyone has at one time or another used a magnifying lens to focus the sun's rays upon something which burns—clothing, cigarette, would shavings, paper, etc. The rays which impinge upon the side of the lens facing the sun are gathered together and focused so that they concentrate this energy to a tiny point of light which becomes very bright and very hot on the other side of the lens. When you hold an object at just the correct distance away—where you get the smallest point of light—the heat becomes so intense that the object gets hot and burn.

Next, consider the size of the light source to the size of the optical fibers and you'll begin to see what could be a problem. Recall that the optical fiber core may be only 50 to 130 microns in diameter, and the larger the core the more tendency there is to have a multimode propagation which causes losses. Light rays may enter the fiber through a numerical aperture. The numerical aperture is the sine of the angle associated with the cone of propagation into the fiber. For example, at Bell-Northern Research, light rays enter a modern fiber within a cone which has a half-angle of 10°. In this case, the numerical aperture is said to be sine 10° = .18. A large NA (numerical aperture) means a larger signal, or ray loss, and larger distortion of the intelligence being thus conveyed.

When you think of the size of a LED and the size of an optical fiber, you may ask "How the blazes do the LED rays get into the fiber optical light guide?" Examine Fig. 5-3 for a general illustration of this problem.

If you can back up so that you have the integrated circuit structure of the LED or laser light source, and have some physical method of bringing the end of such a small fiber light guide into the acceptance cone angle for the guide, then perhaps you have it made. The smaller the light source is physically, the better you will be able to capture the light rays which emit from it.

When you do your experimentation (using, perhaps, a LED from Radio-Shack and an optical fiber from Edmund Scientific) you'll be confronted with this problem. You'll probably just wiggle the fiber around, or the LED, or both, until you observe the glow at the fiber optical stand tip, away from the source, which proves that light is passing through it and emerging at the far end. To measure the losses you will have to use some method to determine the emitted energy at the LED surface and compare this with the emitted energy at the far end of the optical fiber. This measurement is a project all by itself.

Let's get back to the subject of lenses. If you can use some lenses to focus the light into the optical fiber and some lenses to magnify the light coming out, then you may better understand what is going on inside, and make better use of the fiber for whatever purposes you might have in mind. Consider two types of lenses or focusing methods. You can use a parabolic mirror behind the light source and place the input end of the fiber precisely at the focal point of the parabola, as illustrated in Fig. 5-4, which gives some parabolic equation forms and the standard form for the type parabola shown. The focal distance from the vertex to point F is always the same as the distance from the vertex line to the directrix line. In this illustration, the focus point is 3.52 squares from the vertex and lies on the x axis. Any light rays coming into the parabola from the right (as you view it) would be reflected by this mirror to the focal point. Thus, it tends to gather quite a large bit of energy (depending on its size) and it concentrates light at that one point. If your fiberoptic light source is broad-

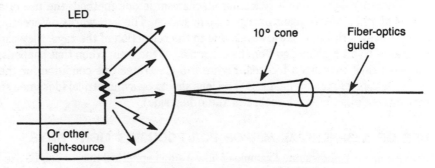

Fig. 5-3. Getting light into a fiberoptic lightguide.

123

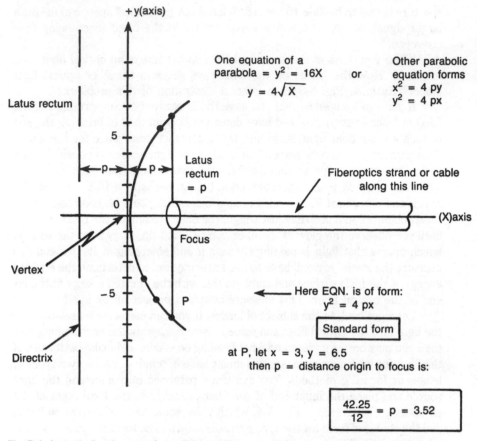

Fig. 5-4. A parabolic mirror can focus light rays like a parabolic antenna focuses radar waves.

beamed, this type of focusing element might be important. The parabola, of course, is three-dimensional and in any direction from the vertex it has this shape as shown. It is, in a way, like a saucer or curved surface which tends to cup the received energy going into it.

Focusing light using a reflecting mechanism is one method; the use of a lens is another way to focus energy rays or waves. The lens works differently, as light passes through it. However, due to the refraction of the rays, they can be caused to hit a given point on the other side of the lens. When that happens, the light rays have been focused. Figure 5-5 illustrates this condition for the thin-lens type of magnifying glass. The lens should be symmetrical (designed so that one side is an exact replica of the other side).

USE OF A SPHERICAL MIRROR FOR FOCUSING LIGHT RAYS

Figure 5-6 shows one example of how a small spherical mirror can be used to focus the rays of light from an object at O to the focal position at F where the

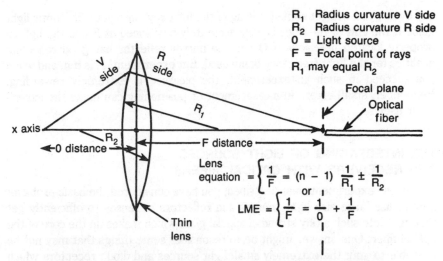

R₁ Radius curvature V side
R₂ Radius curvature R side
O = Light source
F = Focal point of rays
R₁ may equal R₂

Fig. 5-5. A focusing-magnifying lens and the lens-makers equation.

end of the fiberoptics might be located. It is possible to determine the measurements and distances involved by using the equation shown. Just be sure to use the same units (I suggest metric) to find the value of r or the value of F and O.

Normally, when you have an expression in three unknowns you have to eliminate some of them to get a solution. So, how do you do this? One simple way is to select a small spherical object which has good reflective qualities and is shaped as shown, and measure the radius r; this gives you one of the necessary values. Second, you might specify a distance from the vertex to the light source O and then solve for the remaining unknown, the distance from the vertex to F. This may not be the best way but is a method that should work. In

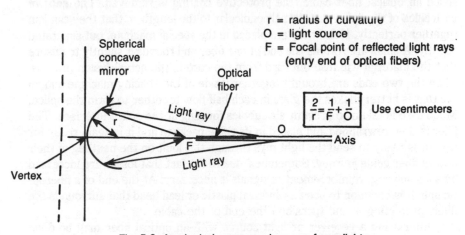

r = radius of curvature
O = light source
F = Focal point of reflected light rays
(entry end of optical fibers)

$$\frac{2}{r} = \frac{1}{F} + \frac{1}{O}$$ Use centimeters

Fig. 5-6. A spherical concave mirror can focus light.

any event, you will get some focusing of the light rays into your fiberoptic light guide if the end of the fiber is very accurately positioned at F and the light is properly located precisely at O and the mirror reflector has good reflecting qualities for the light frequency being used. But experiment! It is fun, and when you succeed in such an experiment, the pleasure is extremely rewarding. Remember, small errors in measurement or positioning can make the experiment useless.

THE INTERFACING OF LIGHT SOURCES AND RECEIVERS WITH OPTICAL FIBERS

In the experiments just described, you have considered the basic problem: how to use large size light sources and reflectors or lenses to efficiently get that light into such a tiny strand of special glass which makes up the core of the optical fiber. One answer might be to reconsider some things that may not be available to you; the extremely small light sources and diode receptors which can be connected to the ends of optical fibers using special splicing and interfacing machines. Once you can make these connections and have these kinds of devices, then transmission of intelligence, data, or whatever using light rays is not to great of a problem.

The proper coding of data becomes a problem, because the data might get lost due to delays in propagation and then changing shapes of pulses. Also, reinforcing the lost energy in transmission is required and as distances involved are relatively short, this may become a problem. Lifetimes of components and accuracy and stability of the light ray frequency must be considered.

So, how can you interface two optical fibers so that the light rays pass through them? Bell-Northern Research has come up with a simple method which is effective in the field. The optical fibers are joined together by a fusion process which is made possible through the use of a special splicing machine developed just for this purpose. What happens is that when it is necessary to splice an optical fiber cable, the protective coating is removed. The ends of both sides of the splice are cut perpendicular to the length so that they can join together perfectly. Next, they are placed in the special machine, but separated slightly. A small arc softens the end of one fiber and rounds it slightly to ensure that the material will fuse outward from the core in the actual fusion process. Then the two ends are brought together inside of the splicing machine and an arc that is hotter makes the glass in each half flow together to form the splice. Much care is taken to prevent air bubbles forming at the splice interface. The fiber is then rewrapped and placed in its cable package and it is again ready for use, or is ready to send the light rays through this joint on the next leg of their optical fiber guide journey. Sometimes, due to the fact that losses are increased by such splicing, reinforcement of signals is necessary. At the end of a fiberoptic line, it is common to use a cylindrical plastic or lead head that surrounds the fiber, protecting it and (possibly) the end of the cable.

Interfacing a receiver or light source with an optical fiber may be done mechanically, or it may be done by properly fusing the fiber and the element into one unit. Also, a jacketed optical fiber can be used to connect an optical fiber to the card edge optical connector from a laser. From the card, a distribution system can be fabricated so that the cable connection to the optical fiber containing the light ray and intelligence can be accomplished in a multiple connecting arrangement.

If you bring in the light and intelligence on one fiber, physical interfacing can be used to connect one optical fiber to a multitude of other fibers and thus send that intelligence over many lines. This is desirable if you are using the light ray for telephonic communications.

SPLICING OPTICAL FIBERS

Four methods of splicing optical fibers are shown in Fig. 5-7A. At the top (A) we see the arc mentioned previously described. A special machine prepares the ends of the fiber, supplies the arc, and then brings the somewhat plastic ends together for bonding molecularly. A strengthening jacket or sleeve covers the splice. In the second illustration (B) you have a pressure bonding of two fibers whose ends have been carefully cut so they are exactly flat and parallel. A metal jacket covers the cladding and plastic covering of the fiber and holds the two ends together. This is not as good as other methods, as strain may separate the fibers. In the third illustration (C) an optical adhesive glue is used to bond the ends of the fibers together, and then the metal jacket covers the splice to give strength to the joint. The fourth method is a multiple-step process using both adhesives and special ferrules to form a strong junction (see Fig. 5-7B).

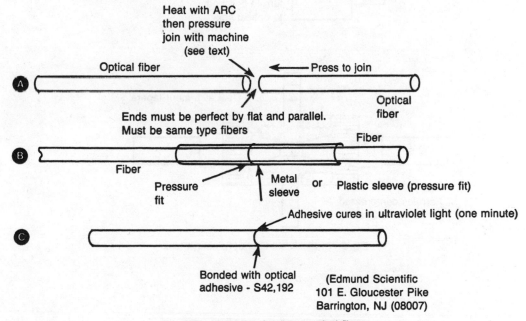

Fig. 5-7A. Methods of splicing optical fibers.

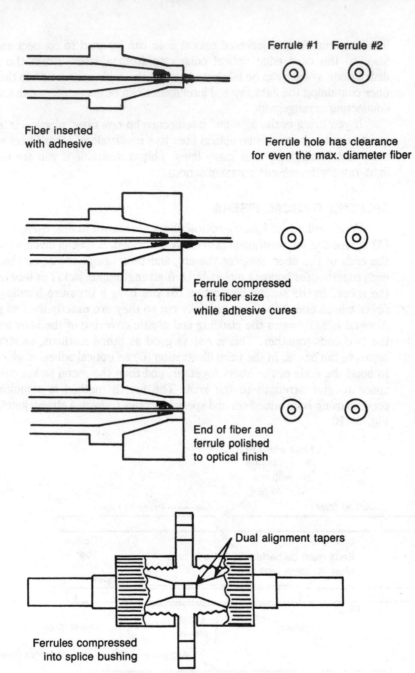

Ferrule #1 Ferrule #2

**Fiber inserted
with adhesive**

**Ferrule hole has clearance
for even the max. diameter fiber**

**Ferrule compressed
to fit fiber size
while adhesive cures**

**End of fiber and
ferrule polished
to optical finish**

Dual alignment tapers

**Ferrules compressed
into splice bushing**

Fig. 5-7B. Multimate fiber-optic connectors.

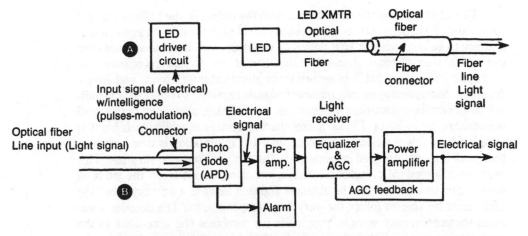

Fig. 5-8. A light-transmitter block diagram (A). A light-receiver block diagram (B).

The following illustrations show (in a very basic sense) how to make an optical fiber transmitter and receiver. (See Fig. 5-8.) A laser transmitter is somewhat more complicated. The laser must be cooled or at least maintained within a certain temperature range to emit the right kind of coherent light. Another simple diagram of a fiberoptic communications system is shown in Fig. 5-9.

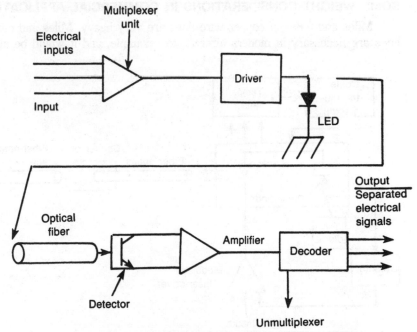

Fig. 5-9. A generalized fiberoptic communications system.

The LED driver is the unit which causes the output of the LED to vary (or cut on and off) at some rate equal or proportional to the combined input signals that are to be converted to light pulses or variations and then transmitted over the optical fiber channel. Remember that one of the advantages of using the fiberoptic channel is that it is secure from electrical interference and eavesdropping. Multiplexing means combining signals together so they are transmitted as a complex modulation; for example, a flute, harp, piano, and horn are all transmitted by radio or TV, or a timesharing system in which each input is sampled for a few fractions of a second and that information is then transmitted, so all lines are sampled in sequence). The sampling must be very fast. This requires a somewhat complex electronic switching circuit to do the job if all information on each line is to be obtained and sent along as a light ray. Also, the LED must be able to follow the very fast sampling units. The decoder separates these frequency units or time units (or combines the time units as the case may require) so that the originals of the inputs are preserved at the output. Some symbology is used for the LED and photodetector which are common in practice and relate to diagrams of photo-optics you may find anywhere.

The output power of a laser light source varies more than an LED when the laser is subjected to temperature changes. This means that a circuit that will monitor temperature (requiring a temperature sensor and an output sensor) must be connected so it causes a proper feedback to stabilize the lasers output. One example is shown in Fig. 5-10.

SOME WEIGHT CONSIDERATIONS IN COMMERCIAL APPLICATIONS

Miles and miles of copper wire lines are very heavy. Miles and miles of lines are necessary in modern aircraft, for example, and it would be nice to

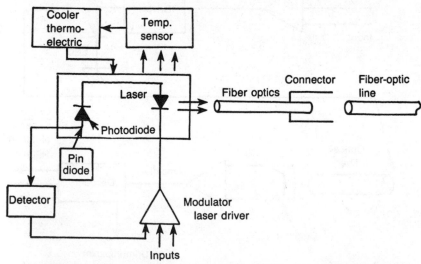

Fig. 5-10. Stabilizing a laser light source transmitter by using temperature control.

reduce this weight so the airplane could be more cost effective. 1.5 pounds of fiberoptics can replace as much as 30 pounds of copper wire in an airplane of the A7 type. Also, the electromagnetic field associated with the passing of current through the copper wires is not present when fiberoptics are used. This gives a quieter atmosphere for the complex instrumentation in many types of airplanes.

At the same time, one of the principal disadvantages of using fiberoptic cables is that they cannot carry direct current. This type of current is commonly used for control and switching applications as well as for power to the various units (amplifiers, equalizers, and sensors) that are part of the optical fiber systems. The way many companies get around this problem now is to include a few pair of copper wires in a fiberoptic cable sheath which may have many optical fiber strands or lines also contained therein.

THE OPTOCOUPLER

For experimental purposes you can get from Radio Shack an optocoupler with a linear op-amp output. Because such a device helps in understanding the use of light as a transmission channel. Let's examine it briefly here. The optocoupler is an integrated circuit unit. It is very small, about $\frac{3}{8}$ inch by $\frac{1}{2}$ inch by $\frac{1}{8}$ inch, with three prongs extending from each side. Inside this hard plastic package are the light-emitting source, the receiving diode element, and a small operational amplifier. The amplifier requires 12 volts dc at 13 mA. The input light diode uses 3.0 volts at around 50 mA.

The operation is basic. When voltages and current are applied to the input diode, it produces light rays inside the structure that go to the receiver diode element. These are then converted by the receiving diode into electrical impulses whose magnitude is dependent upon the intensity of the light impacting the diode element. The receiving diode is then connected to an operational amplifier which amplifies the direct currents and provides an output through one of the pins that is exactly proportional to the output of the receiver diode's output. You can then use this signal for various purposes. Figure 5-11 shows how the connections to this type unit are made.

What is interesting about this unit is that the infrared light rays are emitted from the gallium-arsenide diode. They are coupled physically through space to the receiver diode without any light-guiding path necessary. The diode emits a beam which (for a short distance) provides enough intensity on the receiver diode so that the light-coupling takes place. You can use such an element coupled into an optical fiber in such a manner that its light beam goes down that fiber for a much longer distance before it terminates on a receiver diode which reconverts the light rays back to electrical signals. The amplifier is required to amplify the signal to a usable level.

If the voltage output of the op amp is routed into a circuit which provides a current output exactly proportional to it, then you can use this current to cause another GaAs diode to emit light rays. If it is coupled to another optical fiber, you have the equivalent of a repeating unit (on a small scale) such as is used in telephone communications with optical fibers.

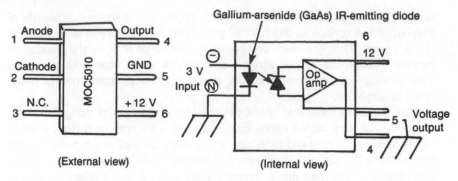

Fig. 5-11. An optocoupler with a linear operational amplifier.

It is current practice to use optoisolators when connecting a computer to outside machines. Although the path between the light emitter and the light receiver is very small, it is an air path through which no electrical connection is made. This means a complete isolation of the light emitter side of the circuit. Thus, any stray voltages, static, or malfunctions that happen in the output circuitry are not coupled back into the output (and into the computers). Often the input side of the circuit is concerned with very small voltages. You can use these to control larger currents and voltages in the output by activating such devices as a triac (voltage controlled switch) with the current from a receiver diode (this voltage can be amplified as necessary by some integrated circuit transistor or transistors in a Darlington-type arrangement). You can find such devices for sale in radio parts stores as an integrated circuit of very small size, but with large capability.

What all this proves is that control functions can be accomplished using a light ray path to convey the control information. This signalling process is relatively slow; hence the light emitter and receiver are relatively slow in response time. When considering the use of transmitter and receiver elements for use with fiberoptic systems, you must think in terms of a very fast, turn-on and turn-off condition for light. If the system transmits fast pulses using the binary code, it must have just as fast a response front as the receiver so that the pulses are reproduced with the sharpest edges possible even though they are being reproduced at a very fast rate because they were so transmitted. In some applications, up to 50 megabits/second might be the digital rate used.

SOME OPTICAL MIXING IDEAS

Basically, all you have to do is to take the light emitted from one optical fiber and reflect it into many fibers in order to get a distribution system. A special coupling device can be used for this purpose that has the proper size and mirror characteristics to provide the reflections in the proper directions. Figure 5-12 might help you visualize this kind of unit. The whole unit is encased, and the mirror must be specially shaped and positioned to get the maximum return of light energy into the multi-strand optical fibers in the

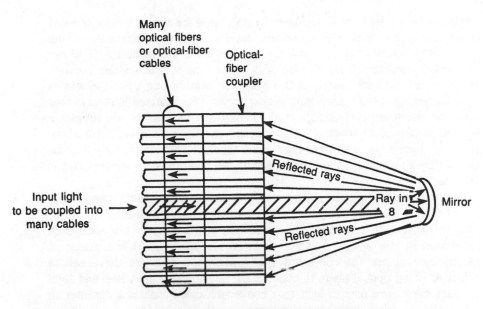

Many optical fibers or optical-fiber cables

Optical-fiber coupler

Reflected rays

Input light to be coupled into many cables →

Ray in 8

Mirror

Reflected rays

Fig. 5-12. A multi-strand coupler for optical rays.

block. As you can see from this illustration, the information contained on the input light rays is now distributed to many channels by this method.

The mixing of signals means that the optical-fiber system must be able to mix different light rays. Perhaps this can be stated more clearly. In a mixing process, the variations of light intensity from dark to full brilliance must then be imposed on other rays going to other channels. You may find that some electronics might be required to do this job.

Let's take a light ray from a fiberoptics line and pass it through an electronic circuit which converts the variations in light intensity into a proportional electrical signal. Amplify that signal and use it to modulate an existing beam of light. Now you have mixed the signals on the original beam with those from the demodulated beam of light.

Some physical devices can vary their transparency to light rays if an applied voltage is placed across them. Some crystals have this property, so they might be used as modulators for a light beam that is the carrier of the mixed signals. Let's start with a separate beam for the carrier. All incoming signals that are to be mixed are converted into electrical pulses. These pulses are mixed to produce one complex electrical signal which is then passed through the crystal to modulate our carrier light beam. Demodulation requires a receiver, a light-to-electrical-current diode and amplifier, and some demodulators or filters to again separate those electrical signals back into each individual channel.

FURTHER USES OF LIGHT RAYS AND OPTICS

I have mentioned ranging and surveying using light rays that can be flashed to a distant point, reflected and timed so that distance can be easily calculated as it is with radar. Unlike radar, light rays are not vulnerable to ECM (electronic countermeasures) or nuclear residue in the atmosphere and so on. You might wonder whether or not the actual effect of atmospherics might not be a prime consideration in the use of light rays and optical devices. Perhaps you have traveled the deserts of the world, or looked along a hot highway at midsummer and have seen a shimmering scene. The reflected light rays that make up vision are certainly affected by the rising heat waves. Sometimes a lens effect is present which tends to magnify or reduce the scenes around us. Although light rays are better than some types of electronics, they may be susceptible to other effects which are bad. You need to be aware that this is possible.

Some sensors operate on the existence or absence of variations in light intensity. It is a simple task to provide a rotating wheel with a small mirror or polished surface that reflects a beam of light into a receiver cell once each revolution of the wheel. A digital counter can easily count the pulses of electricity that are the converted light pulses, and give you the speed of rotation. Using optical fibers to conduct the light rays to the wheel and then conduct the return flash-of-light to a conversion diode enables a designer to locate the sensitive equipment, and computing the equipment, where it functions best.

Light rays can be used to determine colors, and this can be a boon to manufacturers who want a machine which can detect and separate various items that are color coded. It is a well-known fact that light rays going through a lens filter can be restricted or enhanced by that filter. A green filter may pass or inhibit more of the rays in that part of the spectrum than it other colors. Thus, a big output in reflected light to a sensor could mean a green object is present in the view of the light source and the receiver. A restrictive green filter discriminates against the green light rays. Anything green or producing green light only produces the smallest possible electrical signal or maybe no signal at all.

Green, red, ultraviolet, blue, yellow, and other filters are used by camera buffs. Adapting these into the world of sensory optics is already an accomplishment. You can easily conduct an experiment which brings home this idea of the magnitude and type of reflected light. Just use your camera lenses, or a light source and a light meter, and measure the reflected light from various sheets of construction paper available in most grocery stores and pharmacies. Use large sheets, about 18 by 24 inches. Try a variety of colors such as white, green, black, red, orange, and yellow. Always have your light source exactly in the same position with respect to each sheet, and the diode-type detector circuit or meter located in the same position relative to each sheet. If you use an SLR camera, position it in the same position with respect to each sheet. Then make a graph of the meter readings and you can determine the changes in reflectivity for the various surface colors.

To use this idea in other situations, you need a diode-type sensor which is frequency sensitive. Try Radio Shack to find some diodes with the largest output with red, yellow, green, or orange light. Connect your unit so that its electrical output is amplified by a small operational amplifier to a value of 5 volts or so. Use this voltage with some transistors as amplifiers (current types) to operate some small relay of the reed or moving pole type. You can use that to control larger currents.

Another way in which a light-emitting diode and a light-sensing diode are used is to place them opposite each other so they can detect the passage of anything between them. The pulsed light beam announcer is an example of this, and these units are available through most radio parts stores.

The close-proximity diode light source and receiver have a most important application in a shaft encoder and decoder system. In this context, the light is on one side of a disc and the receiver diode is on the opposite side. The disc has holes or slits in it. If the disc rotates from a given position to the right, then counting the pulses tells us the position of the shaft. This is very precise if there is a hole every 10° around the edge of the disc. A computer can keep track of how many pulses are accumulated in a forward direction as the shaft is rotated to move an arm, a tool, or whatever. The computer can then reverse the process by causing the shaft to rotate backward the same number of pulses, or stop at any fractional number of pulses between the maximum number and zero pulses. Thus you can sense shaft positions with a very tiny optical encoder-decoder system very accurately using light rays. There are no physical connections or limitations on the optical encoding-decoding system as far as rotation is concerned.

Using light rays and optical fibers, you can see down deep into machines and see what was not available before without disassembly. You can see what is going on, how parts are wearing, if things need oil, or if parts are broken. Optical fibers can bring out visual signals of correct or incorrect operation to the dashboard of an automobile.

Light rays can also be used to measure the level of fluids. Some commercial and medical applications use this method. Just place a light source on one side of a transparent vessel (glass) so that light rays can shine through when there is no fluid at the light level. Then place a photodiode sensitive to that light color on the other side of the glass container. Connect it through an amplifier to a control valve that is electrically operated. When the fluid is high in the container it reduces the light and even prevents it from passing through the vessel. The photodiode then receives no light and produces a no output. If the fluid level drops the screening effect of the fluid vanishes. Light rays pass through the vessel and hit the photodiode causing a voltage output signal which commands the control valve to operate and refill the vessel until the liquid is again between the light source and the photodiode. It is similar to the operation of a commode, but instead of using an air float to turn the water on and off, you use light rays, electrical signals, and an electrically-operated fluid control valve. In the medical field, the latest technology to accurately control. I.V.'s is to use a pump controlled by a drip sensor. As the sensor detects a drop, it sends a signal to the pump to create a partial vacuum pulling the fluid from the bag at a well controlled rate. What a vast improvement over a nurse who times drops being pulled down only by gravity!

THE BASIC CONTROL ARRANGEMENT
USING LIGHT RAYS AND OPTICAL FIBERS

Whether for a robot (see TAB robotics books, numbers 1071 and 1421), for steering machinery, positioning something, control arrangement, or identifying something by its presence or absence, the basic control arrangement is shown in Fig. 5-13.

The idea is simple. Implement an LED or other light source a light color which matches the photodiodes for maximum output from the reflections. This means that if a photodiode has its highest output with a red color (infrared), use an LED which produces that color as your light source. You can use white light (which has all colors) or other colors (green, blue, violet, yellow) provided you match the source and receivers.

Position the light source so it shines on the object or path and position your photodiodes so they get the reflections from that object. When the reflections are equal in intensity, as determined by the comparison amplifier output, the voltage will be at or near zero. If the object shifts so that the reflected light is stronger on one photodiode than the other, then that diode produces more output. The output from the comparison amplifier is either positive or negative, depending on how you've arranged your circuit. This can be used to state the direction something must move to reestablish balance in reflections. The amount of the comparison amplifier's output can tell the machinery how far to move or how far and how fast to move to keep up with the object.

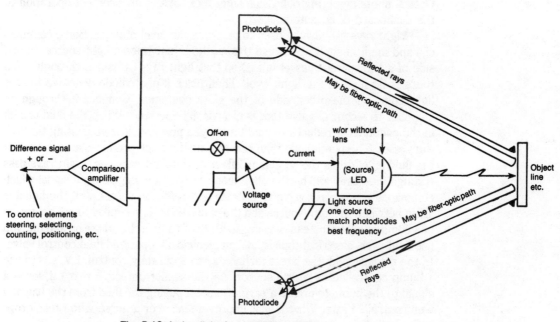

Fig. 5-13. Using light for control purposes.

136

USE	SCHEMATIC	ELEMENTS
Fluorescence spectrofluorescence of liquids, solids, gases also in industrial monitoring liquid chromatography	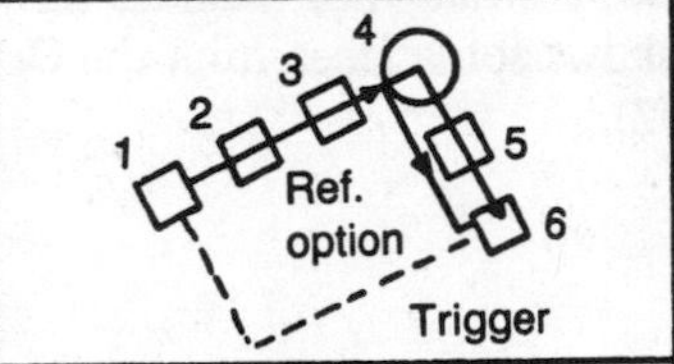	1. Pulsed light source 2. Optical interface 3. Monochromator or filter 4. Sample 5. Monochromator or filter 6. Pulsed light detection system
Diffuse and specular reflection spectroreflection (reflection in daylight!)	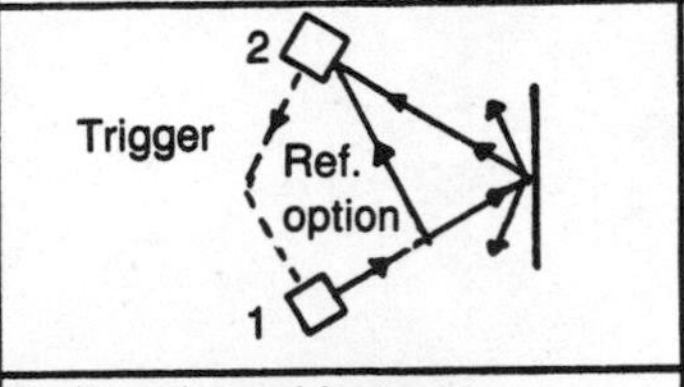	1. Pulsed light source 2. Pulsed light detector
Micro or inaccessible samples	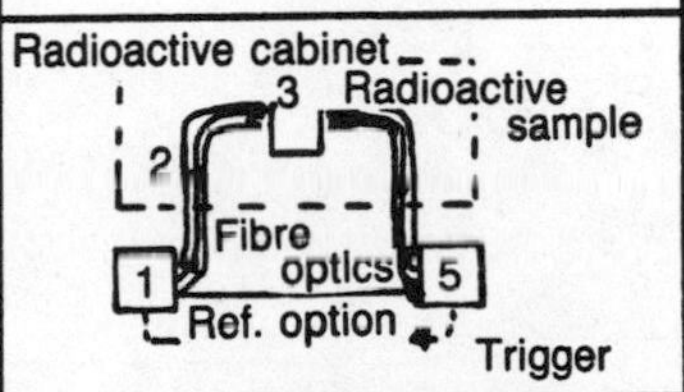	1. Pulsed light source 2. and 4. Fiber optics 5. Pulsed light detector
Absorption, fluorescence + reflection microscopy Immunofluorescence	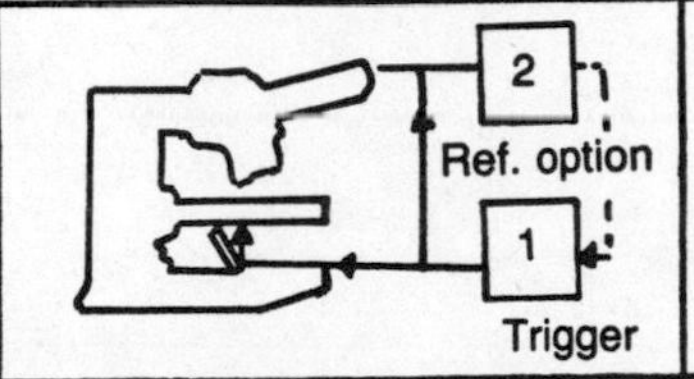	1. Pulsed light source 2. Pulsed light detector
Two channel detection	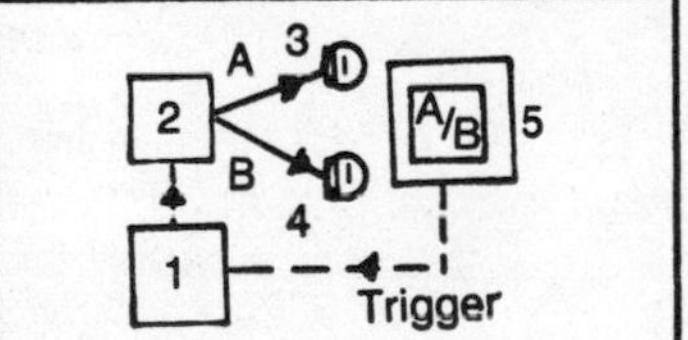	1. Pulsed light source 2. Sample 3. Detector A at wavelength A 4. Detector B at wavelength B 5. Pulsed light detector ratio or 2 channel readout

Fig. 5-14. Some optical systems you can assemble from standard parts (courtesy of Oriel Corporation).

Using this method you can track a line, cause a remotely controlled car to follow a line on a floor, cause a robot's gripper to close on an object, or search for an object. If you want to put your light source on a moving object and photodiodes on another object capable of moving, then the second object can be caused to follow the first by moving it so that its light reception is equal in the two separated photodiodes.

COMMERCIAL MANUFACTURERS OF LIGHT DEVICES

Exploring the possibilities of using light for various purposes offers the potential for lots of fun and profitable systems. If you need optical devices,

Oriel Corporation is one manufacturer which makes them. They have a handsome catalog which shows what they manufacture and how to use their products. Figure 5-14 shows some ideas from the Oriel Corporation of Stamford Connecticut (06902).

Chapter 6

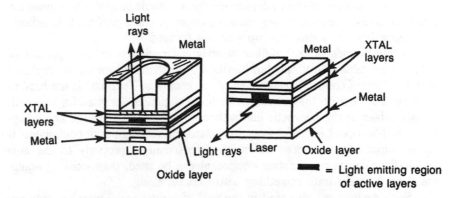

Manufacture and Use of Optical Fibers

Iᴛ ɪꜱ ꜱᴇʟꜰ-ᴇᴠɪᴅᴇɴᴛ ᴛʜᴀᴛ ꜱᴏᴍᴇ ᴋɪɴᴅ ᴏꜰ ɪɴᴘᴜᴛ-ᴏᴜᴛᴘᴜᴛ ᴅᴇᴠɪᴄᴇ ᴍᴜꜱᴛ ʙᴇ ᴜꜱᴇᴅ with optical fibers. Such equipment should enable the user to use existing electronics with this relatively new means of the transmission of signals. This is not a new method; as far back as Thomas Edison's time, some experiments in the use of light rays to convey intelligence were being conducted. Also, Alexander G. Bell did some work with optical systems in his Photophone experiments in 1880. It wasn't until the 1900s that development of the actual optical-fiber light guide, and analysis and development of the theory of how it worked, enabled scientists to come up with useful devices such as the fiber-scope (endoscope) for inspecting dark and dangerous places.

TELEPHONIC COMMUNICATIONS PROBLEMS

Let's consider some aspects of fiberoptics: they do exist, they are most advantageous, they let some things be done better than before (both easier and more economically), they are still being researched and developed. There is an enormous problem involved in telephonic communications (not just the voice part but also the data transmission requirements and the video requirements and such).

Try to imagine what it would take to connect every human being in the civilized world to every other by means of a reliable, reputable, consistent communications system. If that doesn't boggle your mind, imagine what it would take to be able to permit all these people to talk to each other at the same time.

Most people have seen telephone crews down in the hole or up on the pole carefully splicing the countless numbers of specially color-coded cables together so that communications can be established, maintained, or continued to some area. Considering the birth rate and the inevitable increase in demand for more and more communications, data, monitoring, and signaling, it is difficult to imagine how they can keep up with the demand.

A few pounds of optical-fiber strands can replace countless pounds of copper wire cables. Gradually but inevitably, telephone cables will be replaced with this kind of connecting communications channels. That part is not hard to imagine. But how do they connect exisiting equipment designed for use with copper cables to the fiberoptic line without a major design change and equipment modifications? Is it necessary to replace and redesign such equipment? If proper input-output devices which interface (connect) properly to the new optical-fiber cables and existing equipment can be used, then existing equipment can be used until something better comes along.

Such devices are also used in the field of computers. There are tremendous advantages that might be gained by the use of optical-fiber cables between mainframe equipment and terminals. Reduction or elimination of induced radio-frequency interference and the ability to go some places wire cables cannot go due to chemical, temperature, or other problems gives an increase in data handling capability (perhaps as high as ten thousand fold) with less loss in transmission. The computer may need optical communications to speed up the data handling operations. Many scientists say that the travel of electrons over copper, silver, or gold-plated wires is far too slow for the number of operations, calculations, and manipulations currently required, and this doesn't approach the ultimate speed goal of the number of computer operations performed per second: equal to that of the human mind!

THE SIGNALS

You can't interface to anything until you understand the signal transference or conversion involved. In order to know this you must know what kinds of signals are put onto a line. Then you design around the handling of those signals on the type of line or lines used.

Two types of signals are used over wire or optical fibers: pulses and continuous wave signals. Pulses can come in various groupings and codings. They can be sent very fast and enable the use of a large number of channels when they are time multiplexed. (That is, when the signal is transmitted for a certain time duration from one input channel, the channel is switched to another for a short duration.) This process continues until all input channels have been sampled a number of times per second. Each group of pulses transmitted over the connecting line contains a certain amount of intelligence from a particular line. These pulse trains or groups are identified by coding so that they can then be channeled into the corresponding output channels and carry all the essential information from the input line to the output line. Figure 6-1 illustrates this concept.

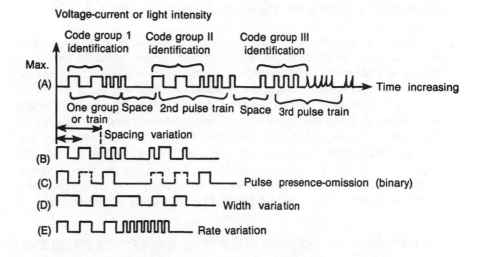

Voltage-current or light intensity

Code group 1 identification Code group II identification Code group III identification

Max.
(A)

One group Space 2nd pulse train Space 3rd pulse train
or train

Time increasing

Spacing variation

(B)

(C) Pulse presence-omission (binary)

(D) Width variation

(E) Rate variation

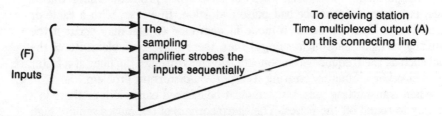

(F)
Inputs

The sampling amplifier strobes the inputs sequentially

To receiving station
Time multiplexed output (A)
on this connecting line

Fig. 6-1. Illustrating the transmission of pulse-trains or groups of pulses.

Figure 6-1 shows how a signal might appear coming over a line when it is time multiplexed (A). Each little group of pulses has some identification arrangement so that the receiving equipment can tell it belongs to a certain group. The receiving equipment then adds up the similarly-coded pulse trains and puts them together to make up the intelligence information (video, binary data, communications, control signals, etc.). At F, all inputs go to a strobing amplifier which is connected to a single line at its output. Thus each group of pulses having whatever kind of coding, such as at B, C, D, E, or some other arrangement, are transmitted and separated timewise during the transmission. So how can intelligent information be gleaned from snatches of data or just samples of information presented at the input? Shannon of Bell Telephone Labs came up with an answer to that. If you sample fast enough, a high rate compared to the intelligence changes, you will never know that the information you get is being sampled at all! Some things like video pulse signals and computer pulse signals may present a different problem because you may not be able to drop even a single pulse from the input line without causing some change in the data output or recovery. For signals such as telephone conversations which are basically continuous wave transmissions, sampling (if done often enough) results in the recovery of the same type sine wave output (by filling in gaps by knowing how the signals are varying in amplitude because of the sampling) without loss of intelligence.

Shannon proved this in his experiments and studies. Humans use a lot of noise that could be omitted, while still getting across to others the exact meaning of what needs to be conveyed. Listen to a conversation and see how many sounds or syllables might be omitted and still not cause a loss of intelligence! Your mind follows the line of thought and so it tends to fill in any blank spaces in actual communication. Shannon showed that in transmitting some signals (such as Morse code) you might miss a letter here and there, but your mind fills in the gaps to make up the proper words. Similarly, while I/O equipment might not send some signals properly, or a few fractional sections by time sampling may be missed, but the receiving equipment (input) can make up what is missing from a knowledge of what was going on before.

TRANSMISSION OF LIGHT SIGNALS (THE DISPERSION PROBLEM)

A single mode optical fiber may not have a bad problem. There will be losses, but there may not be bad pulse distortion. However, with a cable or optical fiber which can be multi-mode in operation (which may occur if the frequency of input gets high enough) then there may some elements of the signal such as the frequency components traveling through the fiber at a faster rate than other elements, creating a dispersion-distortion problem.

When transmitting pulses, there must be a broad bandwidth or there is a tendency to round off the pulses. The sharp corners of the pulses require high frequencies. One way to test audio equipment is to send square wave pulses into it and then evaluate the shape of the audio pulses at the output. If they are sharp and square like the input pulses, then the audio amplifiers are excellent in high frequency response. If the pulses are rounded, the frequency response is not so good. The same idea applies in fiberoptics. If some of the higher frequencies being transmitted are lost due to dispersion and multi-mode operation the pulses which might almost vanish! (Recall that multi-mode means multiple path traveling of the various frequencies and subsequent phase differences that when recombined might cause reinforcement or cancellation of those frequencies). By grading the index of refraction in the optical fiber material, a kind of monomode or single mode transmission of the light ray signals can be accomplished. This way the pulses are preserved in shape and number and the intelligence is faithfully transmitted. Again, remember that the transmission of light rays is just a very high frequency transmission of electromagnetic waves, so the equations and solutions to the light problem can follow the techniques used for radar waves of very high microwave frequencies in following metallic or dielectric guides for these type of waves. (Many texts analyze the transverse electric waves and the transverse magnetic waves associated with the propagation of these radar frequencies, so if you're inclined toward some advanced mathematics I suggest you pursue those as a beginning study.)

To top off this idea, recall that in testing audio equipment with square-wave pulses good flat topped pulses were a good result. That meant that the high-frequency response was good. You can build an amplifier and test its response with square-wave pulses at the input (with a scope and speaker on the output). If good square pulses are produced and they can be maintained as the input frequency from zero to 20 kHz varies you could be very proud indeed. Of course, you might add equalizers and tone controls to make it sound pleasing to your ears!

TESTING OPTICAL FIBERS FOR TRANSMISSION CAPABILITY

It would seem logical to use square-wave pulses of light into optical fibers to test them. You could determine by the shape of the recovered pulses at the output whether the fiber was operating in single or multiple mode by the shape of the pulses. If the effect of the dispersion was large, a group of pulses might appear as shown in Fig. 6-2.

In free space, light rays, like all other electromagnetic waves, travel approximately 300,000,000 meters/second (186,000 miles per second). In a birefringent material used for optical fibers (lithium niobate, for example), the beam divides into two beams polarized with their planes of vibration at right angles to each other. This division occurs when the original light ray is passed through certain crystals such as calcite, quartz, ice, and dolomite. The speed of propagation through the material is different for the two rays, so they tend to doubly refract or have a double bending when passing through the crystals. (Huygens discusses the property of birefringence in his Opera Reliqua.)

Light rays may travel at different velocities through the optical fiber material if they are not coherent and monochromatic. This means that light rays in an optical fiber must be in phase and of the same frequency (common characteristics of laser light). You must carefully choose your input device (the 1 of the I/O) so that it presents the right kind of light ray to the optical fiber for accurate passage through the fiber material with a minimum of dispersion and attenuation. Laser light is one answer, and some LEDs can provide another if they are filtered or generate only a given frequency.

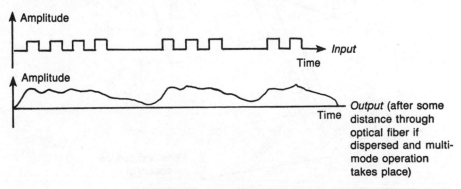

Fig. 6-2. The distortion of light pulses due to inter-fiber dispersion and multimode channeling.

SOME POLARIZED LIGHT EFFECTS

It is difficult to comprehend the concept of polarized light if all you have to use is your imagination, and few of us can get the necessary Nicol prism to experiment with. But you can get some feel for light polarization by spending a few dollars to get a set of polarized sunglasses.

With a pair of polarized sunglasses, try riding down the highway some sunny day and twisting your head slightly. Notice the effect of the reflected light. When the lenses are positioned properly, light reflections are almost cancelled out. A rotation of the lens permits more of the randomly polarized light rays reflected from the highway to pass through them and you may see more light reflections and brightness. Try it!

I recall a college physics experiment in which a grating was used in front of a light. When oriented one way, the grating permitted only the passage of certain polarizations of light. It was this light that you could see. When it was oriented differently you did not see light behind it. So it is with the birefringement concept (or double-refraction concept). The light ray tends to divide inside the crystal and become two rays emerging from the other side. Thus if you look at something from the two-beam side, the input beam is divided and you see a double image of the subject material. Figure 6-3 illustrates this idea.

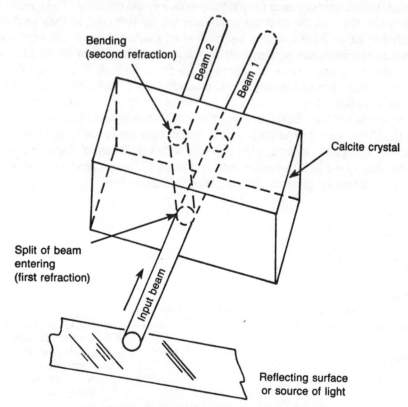

Bending
(second refraction)

Beam 2

Beam 1

Calcite crystal

Split of beam
entering
(first refraction)

Input beam

Reflecting surface
or source of light

Fig. 6-3. The double beam output of a calcite crystal.

144

The commercial world uses a relatively easy method to chop polarized light. It uses a polarized grating. By changing the orientation of the grating, it passes light only when the light matches the grating's polarization. No light passes when the grating is rotated 90°. Rapidly flipping the grating or causing it to change its polarization quickly when a voltage or current is applied is a method of modulating the light rays effectively and producing an off-on switching effect.

OPTICAL TRANSMISSION WINDOWS

Optical windows exist in various materials. What this means is that at certain frequencies waves pass through more readily than at other frequencies. In optical fibers, frequencies about 1.28 micrometers reduce the dispersion effects and the material losses due to wave propagation. They tend to cancel each other and create windows (meaning a really good transmission of those frequencies). Some experiments show that higher frequencies than infrared result in better transmission characteristics and improve the quality of transmission over longer distances than are now common.

The word window related to transmission is not a new concept. In submarine systems, much experimentation has occured to find out what audio frequencies (SONAR) can best be transmitted through the sea water with the least losses. Some frequencies are much better than others. They permit much longer ranging with less power expended by the transducers (energizing devices). The frequency must be kept relatively constant within a window.

You may be familiar with the concept of a rocket-launching window, a particular time when the planets, the sun, and moon are all in a proper position to get a space vehicle into space for rendezvous with some far distant planet. Missing that window (or launch time) means that a long delay might result before an opportune time again arises for such a launch. What you want is the proper window for light rays to be presented to your I/O devices for the best communication, control, and computation. This means that the out (O) devices must match the optical fiber for transmission of the proper light frequencies at the right intensity, with the proper entry angle and with the proper polarization. The (I) input device must be designed to accept that polarization efficiently to insure the most reliable and cost-effective type of system possible.

OPTICAL FIBER CONNECTOR UNITS

As with any relatively new development, there are various devices that can be used with the new developments that are not widely known. Perhaps the optical fiber connectors made by Seiko Instruments are such devices. Certainly you'll think about terminating the ends of optical fibers into something which may then be connected to an input or output device. This connector must be strong and secure to provide the physical coupling necessary for proper light transference.

Figure 6-4 shows the generalized concept of this connector. It is similar to a regular coaxial type of connector for rf coaxial line of the flexible type. But

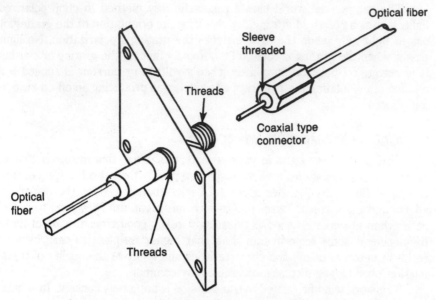

Fig. 6-4. A plug-and-ferrule connector for optical fibers.

there is a difference. The connector of Fig. 6-4 must provide a flat pressure connection to the transparent windows of the sections (even though a pin type insert is used) so that the light rays can easily pass through this fitting. Seiko claims that this is accomplished in this type of fitting.

The unit consists of a plug ferrule, receptacles, and a fiber polisher for single-mode and multimode signal transmission applications. The input strand of fiber is the PVC covered type of 3 mm diameter and is a single fiber only. It is model SAP-2 which has a connection loss of only about 0.5 plus or minus 0.2 dB when in the single-mode transmission state using a 10 micrometer core optical fiber operating in the 1.3 micrometer wavelength region. In addition to the SAP-2 plug, a high-precision ceramic plug ferrule Model SF-1 is required.

If you are considering experimenting with optical fibers and would like to join several lengths together to cover a given distance, then perhaps the plug and ferrule arrangement shown in Fig. 6-4 will be of value. It offers some flexibility in assembling your optical fiber lines. Although you may find some modifications necessary in order to accommodate fibers of other than the stated dimensions (so you can experiment with other than 1.3 micrometer wavelengths) this can probably be accomplished with some advice and assistance from Seiko Instruments or another company such as Oriel Corporation.

THE OPTOISOLATOR AS PART OF THE I/O DEVICE

The opto-isolator can be considered an output device, as seen in Fig. 6-5 which shows how an LED (or laser) as part of an integrated circuit package is connected to some device such as a communications system or computer,

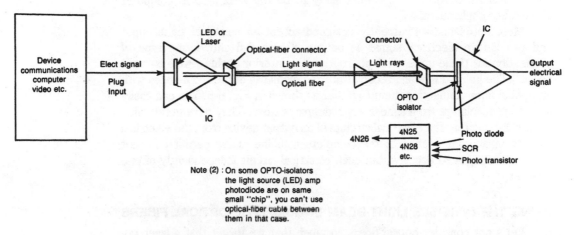

Note (2) : On some OPTO-isolators
the light source (LED) amp
photodiode are on same
small "chip", you can't use
optical-fiber cable between
them in that case.

Fig 6-5. Using opto-isolators as output devices.

which normally has an electrical output that can be adjusted to the proper level of voltage to operate the LED (or laser). The LED (or laser) produces a light intensity proportional to the signal level of the IC's input. This can be an off-on type of signal (binary, as from computers) or a modulation type (varying intensity in amplitude) as from a voice or other type communications set. The light reproduces the variation in that input electrical signal.

You must have a connector attached to the IC to accommodate some type of connection to the optical fiber line. A plug and ferrule connector might be ideal. Next, you must have some kind of optical fiber which has the proper core type (size and index of refraction) to accommodate the desired mode of transmission, single mode or multi-mode. It must also accommodate the frequency of light obtained from the output device shown. It's not as simple as finding any kind of optical fiber, gluing the ends together, and then connecting it to a light source and a photodiode, is it?

The far end of the line or fiber is connected through another connector to the input device that must be some kind of light-to-electricity conversion unit. It might be an SCR if you are going to control something. It might be a photodiode if you want a linear voltage output as we would want from a modulated input. It might be a kind of photo-transistor if all you need is a signal when light is present and no signal when light is not present. Your radio-parts dealer can show you other types of elements or you can find them by examining catalogs from the various manufacturers.

Figure 6-5 shows what could be a connection from a computer to some device or from some device to a computer. With the increasing use of computer-controlled robots (see the TAB books numbers 1071 and 1421) and numerical controlled machines, such connections between and among the various parts of the system seem very likely. Induced electrical interference is greatly reduced using optical fibers in the connecting-communications and data-transmission roles. This seems likely to be the next step in computer system accomplishments.

Most opto-isolators currently designed accept an electrical signal input and provide an electrical signal as output. Figure 6-6 shows one type of connection for these units. With fiberoptics, the interior of this device is "split apart." The light source and receiver are the elements of the I/O devices with an optical fiber connecting them together as shown in Fig. 6-5. You can easily obtain opto-isolators from almost any radio parts store. They have great value in actually isolating the electrical circuits of an output device from the electrical circuits of another section of the complete machine. They permit different levels of voltages to occur within each electrical circuit independently of one another.

USING THE INVISIBLE LIGHT BEAM UNGUIDED BY OPTICAL FIBERS

Let's not consider optical fibers so much that we forget that a laser ray can be propagated in a pencil thin line through space and be reflected back to some source. No optical fibers are necessarily used in this accomplishment. This technique is used in some surveying and leveling instruments, in ranging systems for the military, and in some guided weapons. One company near the missile range at White Sands Proving grounds in New Mexico (Energy Optics of Las Cruces) has developed a ray gun that uses an invisible beam of light to read utility meters from several hundred feet from the meter. This should be no surprise. In supermarkets there are countless bar readers used by checkers to record the price and type of product bought. Checkers run the bar graphs over a special grid which uses light reflections from a small laser to an

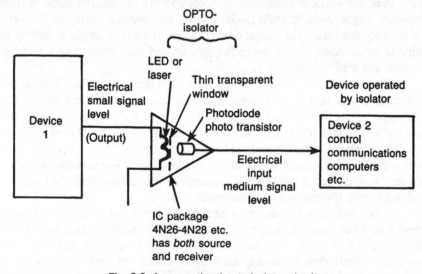

Fig. 6-6. A conventional opto-isolator circuit.

148

optical-electrical conversion unit to produce electrical pulses of wide or narrow width and spacing. These can then be interpreted by microprocessors as binary signals and thus converted into electrical voltages to cause various LEDs to glow brightly on the coat display screen on the counter. The device records what was sold, date of sale, cost of item, quantity of item, and stores it for later use. Let us now examine the concept of one of these bar readers and find out how it works.

HOW A LIGHT READS A BAR GRAPH

What would be required to design such a system? First, it must be able to read the bar graph. Figure 6-7 shows such a graph and the basics of our reading system. Because the graph may take up considerable space, it must be compressed; this means the lines must be narrow and the spacings between them very small. The light beam must be small enough to distinguish between the thin lines and narrow spaces accurately. A laser beam is the answer.

Second, the device must be able to flash over all elements of the bar graph in the small time it appears within the viewing range of the laser beam. Two ideas come to mind. A person might drag the package over the window and thus provide the motion necessary to move the bar graph past a fixed beam position, or the beam itself might move and scan the window in some split

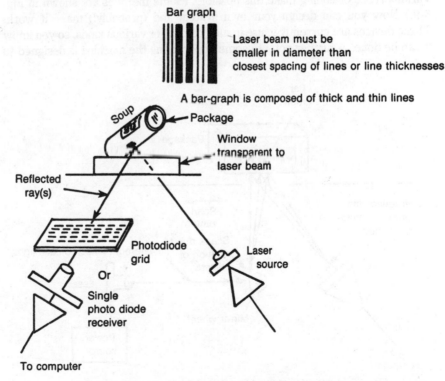

Fig. 6-7. A bar graph and a means for reading it using light rays.

second of time. In the second case it really doesn't matter whether the package moves or not. The moving beam scans so rapidly that the package appears to be standing still even though it is being moved by human hands.

The beam of light will be reflected from the bar graph on the package. It must hit a receiver photodiode so that an electrical signal can be generated by that reflection. When the beam passes over the marked line of the graph the reflection will be of low intensity; thus, the electrical signal will be small. As the beam hits the spacing between dark or black lines, the light reflection will increase and a large electrical signal will result. All you have to do is to devise some way to ensure that the reflection will always hit the photodiode (or photodiodes if you use a bank of them) even though the beam is moving. One method is to use a revolving mirror system to scan. Its own geometry presents the beam to the bar graph at such angles that even though it hits at different physical lengths along the graph, the reflections all come back to a given fixed location (see Fig. 6-8).

Next, you want to send the electrical output of the photodiodes to a computer to make a number that identifies the package as to size, content, and cost. The computer then memorizes the fact that one such package has gone over the window and has been sold. It keeps this data for reordering and inventory. At the same time, it selects the cost and display that for the purchaser, so it must convert the electrical signals into binary numbers. Various types of coding make this possible. (Some methods are shown in Fig. 6-9.) Now you can design your own system and (probably!) make it work. These devices are currently in use in many stores of various kinds, so you know it can be done. A further note: In some applications the machine is designed to

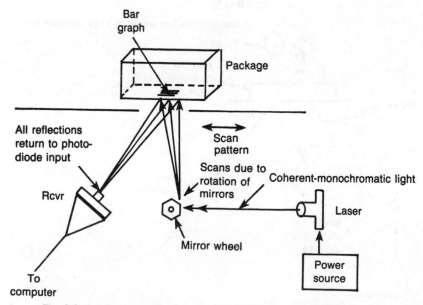

Fig. 6-8. A scanning method to return reflections to a fixed point.

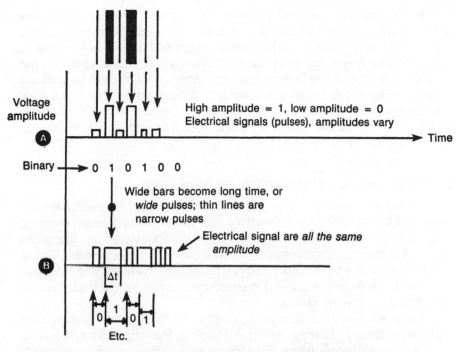

Voltage amplitude

(A)

High amplitude = 1, low amplitude = 0
Electrical signals (pulses), amplitudes vary

Time

Binary → 0 1 0 1 0 0

Wide bars become long time, or
wide pulses; thin lines are
narrow pulses

Electrical signal are *all the same*
amplitude

(B)

Δt

0 1 0 1

Etc.

Fig. 6-9. Coding of light reflections to get binary representations.

give an alarm signal (not a "beep" but a deep sounding "boo") if the package graph has not been read for some reason. It could be because the graph was not flatly presented to the window and therefore could not be scanned. It could have been a fold-over on the package obscuring the graph, and the computer knows when it has not scanned a full bar graph.

How does this fit into fiberoptics? Well, as suggested previously, reading a utility meter from a distance using a laser beam now is a reality (assuming the meter has a kind of graph made up of all its dials that can be read in this fashion). That instrument can use fiberoptics to convey the light beam into the gun that may be pointed toward the meter. Another section of fiberoptics can be used to return the flash of reflected light to a detector deep inside some complex equipment carried on a shoulder or in a car. So, fiberoptics can be involved in this new development and in countless other developments in the future.

Going back to Fig. 6-9, you need to consider some aspects of this illustration. First, in A, the amplitude of the voltage output is what determines whether the computer recognizes a 1 or a 0 from the pulse. It works all right, but this system is subject to interference. If a spurious signal gets into the system, it could easily destroy the voltage levels and cause great confusion to the computer. It could make a 0 appear like a 1 by increasing the voltage output pulse level, and so on.

B of Figure 6-9 shows how this might be overcome. By keeping the pulses at the amplitude (limiting them) the effect of interference is greatly reduced. Then, if the magnitude of the voltage or length of time the pulse exists is stretched slightly you can make use of a pulse width (duration) code that has constant spacing and can be very accurate. Other codes that can be used are pulse-presence-and-omission, pulse-position (in time) with respect to a fixed reference pulse for each pulse train, and graph reading. Perhaps other methods which require a different type of bar graph exist. The field is open, so let your imagination run wild!

APPLICATIONS OF FIBEROPTICS TO ROBOTICS

Robots may use fiberoptics in their extremities and for seeing applications. The use of optical fibers to convey light reflections back to some central location where integrated circuitry can convey the light signals and intensities into usable electrical signals seems almost a mandatory requirement. Optical fibers are used in a robot's hand, and you can imagine them running to various locations in the fingers or grippers in order to locate objects that the robot is to take hold of, move, or pick up.

Sumitomo Electric Industries of Japan has developed a most human robot with perceptions of hearing, speaking, learning, and with arms and legs to approximate the human figure. This humanized robot or android understands human voice commends to such an extent that you can tell it to go pick up something, carry it somewhere else, and place it there. It obeys your command. It is small unit (shades of R2D2!) only 36 inches by 20 inches wide by 39 inches deep.

The information that seems most pertinent is the fact that a description of the robot says that it has two moveable eyes, and that they are made of some 300,000 optical fibers. This permits the robot to distinguish through image recognition technology (see TAB book number 1421) the shapes and sizes of various objects, and perhaps even the colors. Also, the announcement states that the robot has some optical character-reader capability in its legs which permit it to read written messages, and follow various paths as it moves and avoid obstacles.

The state of the robotics art is advancing. The eyes of this robot (if everything stated is anywhere near factual) mean a breakthrough in object recognition by a machine of this type. That optical fibers are used to permit a grid of light pixels (necessary to give the robot some vision capability). This grid might have as many as 300,000 elements, quite an accomplishment. Second, the use of optical fiber strands to connect from the moving eyeball (or light scanning unit, as it might be more properly described) to a microprocessor which ingests the various light reflections to identify the patterns as objects, places, and spaces, seems somewhat marvelous.

Let's discuss the control problem. Imagine what it might be like to design a servo-system to move such a scanner. This small device would use some of the reflected rays to govern just where the eyeballs would point or focus. Then too there is the concept of lenses, perhaps a single lens for the whole grid or possibly one for each fiber (and recall that these are hair-thin in size and each might have its own lens built on to its seeing end). Also consider the multitude of micro-small connectors necessary to connect the distant ends of the fibers to the microprocessor unit.

It is not hard to imagine that a fiberoptic cable might contain a multitude of fibers and be relatively small in size, flexible, and quite easy to handy physically. The scanner eyeball would not have to be physically large. But since this unit (like some others attached to a robot) might have to be moved anywhere within at least a hemisphere, you've got a control design problem of no small magnitude.

Here is another use of optical fibers. This field of automation, robotics, and numerical machines is expanding at breakneck speed in the design of seeing devices for these machines so that they can be given instructions that a person could follow. "See that tool over there? Get it, and fasten this bolt."

The day or wire connections between various electronic solid-state devices may be approaching its end. In addition to printed circuit boards, new boards have fiberoptic cables and lines built in to send data and signals throughout the device. The use of metallic electron roads in electronics may begin to reduce and even approach the vanishing point. Learn and experiment now, so when the time comes to confront this new technology you won't be caught napping.

Figure 6-10 shows how the eyes of a robot might be fabricated. A shows a round eye and B shows more machine-like eyes that are a flat segment in a Cardan suspension system driven by two small servo motors and their associated electronic systems. Each type of eye may or may not have lens arrangements in front of the multitude of fiberoptic ends which pick up the light rays from the reflections and convey them to the microprocessor located somewhere in the body.

USING CAMERA DEVELOPMENTS IN FIBEROPTIC APPLICATIONS

Moving the robotic eye so it can see things requires using servo systems of extremely small size and great precision. The field of photography offers a solution to this problem. A Honeywell development (from the Honeywell Visutronics Group of Denver, Colorado) is a small through-the-lens focusing system that is automatic and autonomous. It was designed for SLR cameras and uses two chips: an electronic eye and an electronic interface to its microcomputer. The light sensor chip is designed so that it produces two signals as output that are in phase electronically only when the camera lens is focused. Here's the clincher for our robotics application: This tiny microcomputer determines the distance and direction (in or out) to move the lens to bring the object into focus using a very tiny servo-system of great precision. This system has an 0.05 millimeter focal distance and its accuracy equals human performance (so they claim) for high contrast scenes, and is better than human performance with low-contrast scenes.

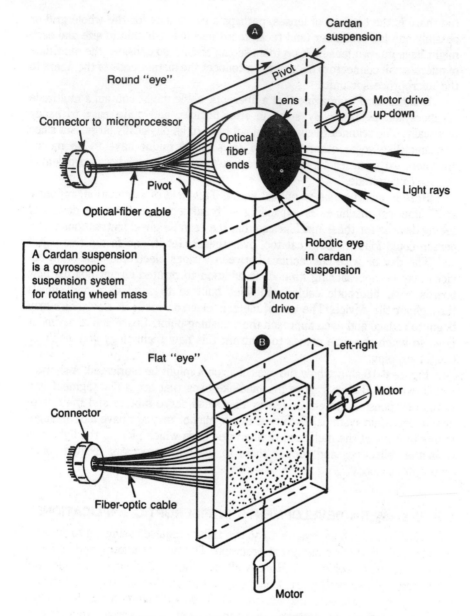

A Cardan suspension is a gyroscopic suspension system for rotating wheel mass

Fig. 6-10. Japanese robotic eyes using fiberoptics.

For a robotic vision system, all you have to add is identity of the light and dark objects seen, or the color geometry of the scenes and objects that the robot might look at. Then it can identify what it is seeing and tell where it is and how it is oriented. There will be much use of robotic vision in forthcoming machines to make them more adaptable to our needs and desires. Fiberoptics will play a most important part in these developments.

MANUFACTURE OF OPTICAL FIBERS

If you want to make a copper wire, you take an ingot of copper, heat it, and start squeezing it down. As it begins to get longer and longer, you keep it semi-round, and finally you push a heated end through a small hole in a metal die and pull or draw it through that hole to make it come out the size you want it to be. Then it cools, you wind it on a reel, and you have your wire. If you want it covered with insulation you run it through a coating bath before you wind it up for use.

Making an optical fiber is similar. Heat some sand, silica, and perhaps a few other chemicals until they are molten. Stir them to get a good even mixture. Then you form the glass rod much like you formed the copper wire. In the drawing process with glass, you may have to add heat to keep the material the right plasticity so it draws properly into those hair-sized optical fibers.

The various types of optical fibers are drawn out on machines that are all basically the same. Some very refractory materials are drawn from solid rods or (as they are called) preforms but the softer glass structures that are fluid at temperatures of only about 1200° centigrade might be drawn directly from special containers or crucibles. What happens is that the glass is put into the machine which feeds it to a heating source. After the glass is softened by the heat and becomes plastic, it is pulled down through the sizing die to give it the proper diameter. Then it goes on down to a wind-up drum where it can be stored neatly and efficiently. (Recall that glass loses heat very quickly.)

If you have ever watched a glassblower at work and marveled at his skill, you remember how he used the heat of his blast furnace to melt the silica, sand, and chemicals into a mass of molten glass. The color of the glass is determined by the chemicals and the type of glass-sand used. The glassblower often stops to reheat the glass and then with quick decisive chops of a cutting tool he lops off unwanted segments as the glass object forms under his expert blowing and shaping. He can add to the glass by fusing new hot glass sections to the base structure and presently he has a vase, swan, bull, or whatever, which is at once delicate, amazing, and beautiful to behold.

So it is with our optical fiber drawing machine. It simply adds heat at the stretching place so the glass fiber becomes plastic and can be drawn out into its tiny diameter. Some machines monitor the fiber size to ensure that just the right diameter is maintained. Also the coating of the core structure may be accomplished as the fiber (correctly sized) is drawn down toward its roll-up drum. Coating helps to preserve the fiber's tensile strength.

The glass used in optical fibers is softened at some 800° to 1200° centigrade. This range of temperatures must be provided to let the fiber be drawn into a usable form. In a method called the double crucible or double-bushing method, the core glass is placed in an inner crucible which has a hole in its bottom and the cladding glass or material is placed in the outer crucible (which has a hole concentric with the inner crucible's hole). As you'd imagine, when the inner glass starts to flow, the cladding glass also starts to flow and it wraps around and fuses to the core glass. The cladding material has a different index of refraction from the core and is made to become a part of the optical fiber strand. Using this method with various types of physical arrangements, cores with graded indices of refraction can be obtained. This can result in single-mode transmission and low dispersion loss of the light rays through the glass material. The two crucibles are fed with preformed rods of glass which have the required chemical characteristics for proper indices of refraction as desired by the user. It is an automatic process once the machines are started and placed into operation. The cladded glass fiber is wound on a storage drum for shipment and use.

Some types of optical fibers are drawn from specially prepared solid-glass rods. Temperatures over 2000° centigrade are required to soften the high silica-content material to a plastic enough state so it can be drawn into the small diameter fiber. This high temperature requirement may require special heating devices like induction heating furnaces. When directly heating the rods, special CO_2 lasers have been used. Using lasers gives a fast temperature response (quick heating) and the ability to draw out the fibers in almost any atmosphere.

There are multi-component materials used to make fibers; for example, the soda-lime silicate and sodium borosilicate glasses. These are prepared by melting silica (that makes the glass) and adding some of the other modifying compounds to the melt. The additives may be carbonates or nitrates which decompose to form the oxides in the melt. It is very important that only very small concentrations of impurities be present in the resultant optical fiber if it is to have a lot-loss light transmission capability. This has to be true especially in that region of color or wavelength that is to be sent over the fiber. To ensure the adequate purity of the fiber, the materials themselves must be pure. This requires sampling during preparation to ensure that only the purest kind of materials are used.

In addition to using pure compounds to make optical fibers, the methods used to melt the glass must also ensure that the fiber is of adequate quality. This means it must be chemically and optically homogenous and free from bubbles and inclusion. (An inclusion is a fracture in the glass structure.) To get the required homogenity (mixing of the chemicals evenly and thoroughly), the melt is stirred or a stream of gas is passed through the melt during the melting phase of manufacture.

In one example of the manufacture of an optical fiber, a premixed silica, sodium carbonate, and calcium carbonate of high purity was melted down in platinum crucibles using rf power (like a microwave oven) to heat the crucible. After the substances were melted, they were stirred with a platinum paddle at 1500° centigrade. The melt was then permitted to soak at this temperature (simmer) until large bubbles floated to the surface and burst. Finally, more stirring was done with the special paddles until the required homogeneity and optical purity were achieved. The temperature was then reduced some 200° centigrade and again a soaking time was used to permit ultrasmall bubbles to be absorbed into the melt. The whole mix was permitted to cool down to about room temperature and annealed (heated and cooled) over a time period of some 20 to 24 hours. If there are high losses in light transmission, it is because contaminations get into the material or the material is not optically pure. Thus, only the purest materials are used and the utmost care is taken at all times in the manufacture of optical fibers.

RF power is used to melt the materials used in optical fibers. This leads to fusing of the ends of fibers themselves so that a continuous line can be made that will convey the thousands of conversations which companies like Bell Telephone are interested in developing. The electrical conductivity of the glass used when it is in the molten state is high enough so that rf power can couple directly into the molten glass material and heat it. Another interesting aspect of manufacture is that if the melt is cooled quickly by water or similar means after mixing is completed, then the melt does not tend to absorb impurities from the crucible walls and thus tends to have the same purity standard as the materials from which it was prepared. Some types of fibers have losses of 27 to 50 dB/km when prepared; these losses are at a specified frequency of around 1080 nm. There are many problems in getting the melt to be consistent and smooth and homogeneous. One problem is that if the melt is heated a little more in one location in the crucible than in other locations, convection currents tend to set up in the melt. This makes for bubbles which are not easily detected and removed. It isn't just the simple process of heat and shape as the glass blowers employ. Here the heating must be uniform, the right temperature, the materials of the right purity, and the melt must not be subject (any more than it is possible to prevent) to outside impurities getting into the melt as the fiber glass is formed into strand.

One of the materials used to make optical fibers is sodium borosilicate glass doped with thallium oxide. This has been called the Selfoc fiber. The cladding of this type of fiber is formed from an undoped sodium borosilicate. The technique used is such that as the fibers are drawn (double crucible method, so that the cladding is over the core material) the cladding material is in contact with the core when it is very hot for some period of time. This permits an exchange of ions between the sodium and the thallium which yields a graded refractive index, which allows monomode type of transmission of light. Hence a graded index of refraction can be of great importance. The losses in transmission using this type of fiber have ranged from 14 dB/m to about 20 dB/m when the light frequency was in the 800 nm range.

Another material used in the development of optical fibers is a synthetic material that really is a fused silica. It permits good passage and low attenuation of light in and near the infrared region. The synthetic fibers are prepared from synthetic silica made by a process of vaporizing silicon tetrachloride and subsequent oxidation or hydrolysis. The material produced was very pure. There is always a catch. Because very high temperatures are involved, a special technique called soot deposition using a white soot that is formed when gaseous silicon compounds are decomposed in the presence of water is used. Using this soot in a special process that first caused the soot to be impregnated on the inside of a glass tube, and then drawing the tube into a solid fiber rod results in the production of very efficient, low loss, fiberoptic light guides.

Still another method of manufacture has been used to develop pure, low loss fiberoptic light conductors, the chemical vapor deposition method. In this method a layer (or many layers) of special material is deposited on the inside wall of a tiny glass tube. Then the tube is heated until it becomes plastic enough to collapse on itself and becomes a rod instead of a hollow tube. All that remains is to draw the rod through the proper die-type material until it reduces to the desired size and diameter. In this method, fused quartz has been used in the deposition process. In this process, the core material has to be further doped with some oxides so that it has a higher refractive index than its cladding glass. This is done using some fused silica doped with other oxides. The core was made originally from germanium oxide doped silica. Later, other materials have found consideration as a core material. It is interesting to consider a chart that shows a window and some attenuation characteristics of glass made from many materials (see Fig. 6-11).

The problem of pulse dispersion in an optical fiber has been discussed in earlier pages. Essentially, it means that for some kinds of fibers and for some propagation methods into fibers, the output pulses are not distinguishable (in the worst cases) and are badly distorted in other cases. This is due to the fact that not all of the light frequencies involved in the pulses are transmitted in the same manner, with the same efficiency, and with the same phase or timing. When recombined in the fiber's output, the signal can actually be a mess instead of clear, sharp pulses. Optical fibers in which the refractive index changes gradually (from the cladding toward the center of the core material in a precisely pre-determined manner) have less pulse dispersion than a nonvariable refractive-index type of fiber. In manufacture, it is desirable to be able to vary the refractive index in accord with the scientist's specifications to accomplish the best possible transmission of light. There are machines that can

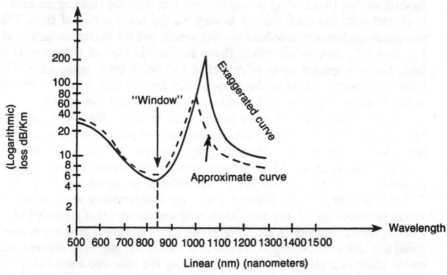

Fig. 6-11. An optical window defined and the attenuation characteristics of one type of light-fiber guide.

accomplish this graded refractive index requirement by using a deposition method and many layers of various types of chemical materials. There are several types of flourinated polymers that have lower refractive indices than the fused silica, and they can be used as cladding materials.

In the manufacturing process, look at the size of the fibers. Much work has been undertaken to develop monomode fibers that have good tensile strength and are large enough in diameter to facilitate splicing. If a telephone company wants to run lines that are miles in length, they want very low losses, good transmission characteristics, and the ability to easily splice the optical-fiber cables and lines. It is not easy to develop an optical fiber which is mono-mode at many light frequencies; in fact, it may not be possible. But some bandwidth is permitted in this mode, such as a band around 1060 nm wavelength. Within this band, the communications capability is almost beyond imagination.

THE BANDWIDTH CAPABILITY OF LIGHT CHANNELS

Let us do some analytical work for a moment. You have heard about the tremendous communications capabilities of the extremely high frequencies involved with light channels. Assume that you have a wavelength of 1060 nm, meaning 1060×10^{-9} meter. (It can be written 1.060×10^{-6} meters or a little over one micrometer.) To find the frequency, divide this into the speed of light (c) which is 300×10^6 meters/seconds.

$$f = \frac{c}{\lambda}$$

λ = wavelength in meters

$$f = 3 \times 10^8/1.060 \times 10^6 = 3 \times 10^{14}/1.060$$

or, for our purposes, we might approximate this as something a little less than 3×10^{14} hertz. That is a big number. If you don't think so, write it out, use a (3) and 14 zeros after it!

A light frequency band might include a few frequencies each side of this center frequency, or say that the band is from 3×10^{14} to 3.5×10^{14} hertz. How many channels of, say, 5,000 hertz plus or minus some 3000 hertz will fit into that higher frequency band of 0.5×10^{14} hertz? A rough mental division produces at least 0.16×10^{14} channels of the audio-spread bandwidth. That's a lot of channels! And that is only the beginning. If you use several light channels, each with the previously indicated capability, you've got enough to handle almost anything now or in the immediate future, including TV and other types of video signals and data signals for computers and control signals. Yes, there is a decided advantage to the use of light frequencies as channels.

You may question the narrow channeling used as an example for the audio channel, 5000 hertz. A much higher range of frequencies might be needed for pulse-type transmissions. You are right, of course. But I ask your indulgence. If you know that, then you also know the capability of light channels perhaps better than most and do not question that capability. For your own satisfaction you might recalculate how many channels you believe are available on a single light channel. I'd like to know your answers.

CONSIDERING THE MANUFACTURE OF OPTICAL FIBERS AND THEIR STRENGTH

You now have some idea how optical fibers are manufactured, how they are made with a single index of refraction core, and how they are made with a variable index of refraction in the core and cladding. Now consider the very practical aspect of manufacture relating to how the fibers are made so they have sufficient strength to be drawn out and still stay all together.

The problem with manufacturing that affects the strength of the fiber is that there are flaws in the manufactured glass of various types. These limit the strength of the fibers as to stress and strain. The closer a flaw is to the fiber surface the more it affects the total strain capability. The literature you might study in connection with the failure of glass structures varies considerably as to the individual testing results. This could be because there is a difference from batch to batch on the fiber material, or because flaws creep into one batch and may not do so to the same extent in other batches or bundles of produced fibers. Even if a glass fiber does not break when in production, it may break later. Surface cracks smaller than the critical size for a brittle fracture may actually get larger under some conditions (for example if water vapor is present). They grow until the fiber actually breaks. This kind of fracture is sometimes called a fatigue factor and may take place after many years of fiber existence.

The strength of an optical fiber is related to its length. The longer the fiber produced, the more likely it is that some flaw is generated or appearing due to circumstances beyond production control that causes fracturing problems later. This is also true with wires that are drawn. If you take a small segment of either one and tie it between two terminals, it might be okay, but take a very long segment and tie it up between terminals or put any strain on it in any way (even if it is lying in an earth-hidden cable) and sooner or later it will develop troubles.

Every effort is made in manufacture to make fibers strong. Every aspect of manufacturing is carefully controlled. In handling and winding of the fibers, care is taken to prevent abrasions which might nick or scar or otherwise affect the optical fiber surface. Sometimes the fibers are coated with some material that adds strength and still does not affect the light ray transmission capability of the fibers. These materials are coated over the cladding. But the most damaging element to the fibers is water vapor which produces fatigue or failure in some reasonably long-term time span. When fibers fail or have to be rejoined, the splicing processes previously mentioned take place. These may be done by machines, men, or both. The method used will be the one that gives the best strength and the least loss of light ray transmission.

Clad-glass fibers do best when they are microscopically straight. They have the least light loss that way. There are no bends in the fiber to start dispersion and multi-mode effects which reduce the effectiveness of the light transmission. As you wind a fiber on a drum surface, that surface may not be perfectly flat. The imperfections on the drum surface (which might be just dust particles, little metal chips of microscopic size, or scratches) can reduce the light transmission capability and produce losses. If the fiber is coated with some kind of protective coating before it is wound, these losses are minimized. The problem is that the optical fibers are so small in diameter that almost anything can bend them in some microscopic manner causing the loss problem. Some companies such as Northern Telecom have designed a special cable arrangement that is (looking at it from the end) a segmented, multiple cloverleaf pattern where each leaf is a slot. Into each slot is placed the optical fiber, or fibers, in a manner that they are not under the stress and strain of the cable material itself, nor are they subject to the microscopically uneven surfaces that might be present against an unprotected optical fiber. In some of the slots are place wires for those control operations which demand such connections, so a cable is a rather larger (but not too large) type of hose device, looking at it unopened. Inside are the works as protected as they can possibly be against all the elements and events which might cause problems in the light ray transmission over the internal fiber light carriers.

Some of the literature, however, suggests that rather than bunching the fibers in cables, they should be put into small groups which are then formed into cables. The two types of cables suggested are the flat type (you are familiar with this ribbon type of wire connector cable used with computers) or into the more conventional circular cable type. In either case, the fibers, which are specially coated, are wound around a special central coated steel wire for strength, in a helical manner. The whole thing has a layer of polymer protecting the interior and the ends are encapsulated in urethane and finally covered with something like polyvinyl-chloride plastic. In the ribbon structure, the fibers have to be separated enough to permit splicing and the ribbon material in which the fibers are embedded must be strong and protective like the steel-core plastic-covered circular cable.

SOME NOTES ON SPLICING OPTICAL FIBERS

When the ends of two optical fibers are to be spliced, the first consideration is to do this in such a manner that excessive losses of light energy will not take place at the splice. This means, for one thing, that the ends of the fibers must be as nearly flat, cut perpendicular to the fiber length as near perfectly as possible. Consider that the fiber is a type of glass and imagine how you would make this cut.

When you cut a piece of glass you use a scoring tool which scratches the surface of the glass. Then you tap the glass somewhat sharply near and on one side of the scoring line while holding the other side flat on a cloth-covered table. It is always an amazement to watch an expert glasscutter at work. He breaks the glass so quickly and cleanly. I've tried it and the result is almost always a jagged fracture which may, or may not, follow the scored line at all! Is the scoring line deep enough, or too deep, or what? Oh well, each to his own trade!

I was not surprised to find that scoring the ends of optical fibers to cause the glass rods to break cleanly and in a flat, perpendicular manner is not always successful. A special machine was devised to do the job. This machine uses a diamond or carbide tip on the scoring tool to make a good clean mark. Manufacturers are able to cut quite effectively but the preparation of multiple fiber ends is another matter. Cutting them in the required flat manner, then polishing them and/or grinding them down to a perfect fit is quite a task.

There is splicing liquid which can be used to glue the ends of fibers together, but this requires some doing because you have to align the ends of the fibers which are to be glued, compress them together tightly until the glue sets, and then cover the joint with some kind of tension-protective jacket. If you can align the fibers and properly glue them together with an index of refraction-matching liquid glue, then you might not have losses over some 0.045 dB for some types of fibers; notably the high silica multimode low-loss types.

Mechanical connectors may be either circular or V types. They butt-join the fibers together. A pressure fitting is tightened to continue to press the ends together so the light rays can pass through this joint. Some say there is an advantage to this type of splice in that you can pull out and push in the fibers to make the connections any time you choose (like pushing a copper wire into a connector and pulling it out when you want to change the circuit arrangement). If you are not careful to make a perfect butt-joint fit each time, you have severe losses or perhaps no transmission of light rays through your connection at all.

There is another method. You take a bit of plastic and groove it so the fibers will lie in those grooves. Then glue a top plastic piece in place, and the joint is secure, solid and cover-protected. When this is done properly, a very good continuation of the light transmission capability is maintained; otherwise you might have to have a magnifying glass or microscope to make sure the fibers are properly joined together. Remember that fibers are extremely small hair-like strands of glass, even with their cladding and some plastic coat covering.

To make a splice requires that the fibers be prepared correctly. They must be aligned so they mate properly. They must be somehow connected together with an index-matching glue, or physically as in fusing with heat in such a way that the integrity of the fiber is maintained even though the glass is heated to a molten state. The index of refraction must be maintained with the joining compound, if any is used. That is no small accomplishment!

HISTORICAL USES OF OPTICAL FIBERS

As far back as 1977, an installation in Chicago by the Bell Telephone System connected two central offices and a customer location in the Brunswick Building to demonstrate the capability of optical fibers in communications with voice, video, and computerized data signals. Lasers, LEDs, and avalanche photodiodes were used to transmit light rays and convert them back to electrically-usable signals. The system was so successful that an effort to use optical fibers almost everywhere began at this time. Bell-Northern Research is doing a fantastic job in the development of this type of device in preparation for the jobs ahead in conveying intelligence from one place to another.

There are some concepts which make the reason for this excitement more understandable. It has been said that the Gutenberg Bible can be transmitted over wires completely in one hour. That same amount of material can be transmitted over an optical fiber system in one second. Word comes from the Bell Telephone laboratories that special optical fibers can now be produced as easily (with much better quality) than ordinary glass fibers. These are called multimode fibers. The glass fibers are just some 49 microns in diameter (25,600 microns to an inch). Also, the single-mode fibers are merely 5 microns in diameter — less than the diameter of a human hair. Pull a hair out of your head and examine it. You are looking at something about the size of an optical fiber. You'll then realize how difficult the problem may be to get light rays into the end of this fiber and to get the light out into a proper detector device. You'll begin to understand why microscopes are used when thinking of working with such fibers, and why there are difficulties in splicing and handling them.

Some 37,000 miles of optical fibers have been installed by the telephone companies and are now being used! These fibers are the multi-mode type and thus are slightly larger in diameter than the single-mode type. When you go from a few microns to several microns (i.e., from single-mode to multi-mode) it is not a big size jump for humans.

The first attempt to use light frequencies as data channels was to use them with the so-called DS3 rate of digital communication which, in North America, operates at about 44.8 megabits/second and is used by the telephone companies in their trunk cable connections. If the rates of transmission are lower, then copper cables do the job well. But the effort is made to develop optical fibers so they can be used with data busing. Bell Telephone Labs has demonstrated a capability to transmit as high as 480 million bits over a mono-mode optical fiber system over 50 miles in length, without repeaters or amplifiers. This tells us something about the perfection of splicing methods used in this long cable!

Some other problems associated with optical fibers are that the signaling techniques using binary procedures make it difficult to design lasers, LEDs, and photodiodes of various types that can follow the extremely fast pulsing rates used (especially when transmitting the megabit rates). Not only do light sources and receivers have to be able to create and accept the pulses, they also have to be able to do it while retaining the pulse shapes, and differentiating among the pulse trains or groups that make up the various kinds of intelligence being transmitted. Thus is no trivial task!

Bell-Northern Research, Ltd. has transmitted digital pulses at the DS1 rate (1.544 megabits/sec) over trunk lines as long ago as 1962. This conglomerate found that the following requirements have to be met by interfacing procedures, devices, and techniques. This must be done electronically in order for optical fibers to actually be used in communications in the North American digital world.

Designation	Digital Rate	Channels VF Capacity
DS1	1.544 Mb/sec	24
DS1C	3.152 Mb/sec	48
DS2	6.312 Mb/sec	96
DS3	44.736 Mb/sec	672
DS4	274.176 Mb/sec	4032

If the optical fiber systems can operate at these rates, they can be interfaced with proper devices to the current telephonic communications systems. Of course, there is always the question of cost-effectiveness of the optical fibers rather than coaxial cables, copper wires, radio channels etc. That has to be determined, and then the choice may be based on the cost effectiveness figures rather than on any technical or manufacturing or maintenance conditions or developments. You can say that cost-effectiveness includes such things as these and you'd be right. But there are other considerations in a cost effectiveness study.

In laboratory experiments, wavelengths as low as 1500 nm have been used very successfully for transmission over monomode fibers. The transmissions have been over very good distances with low enough loss that repeaters were not required.

Because repeaterless spans might be used over some distances—up to 12 kilometers—that using a 850 nm wavelength with laser light sources and APD detectors that 60 to 70 percent of the urban digital spans could be repeaterless. Optical fibers used might be only 25 mm in diameter. Since they are very small and lightweight (even when properly packaged with cladding, plastic wrapping, and in cables) they can be placed on reels which contain much longer spans than an equivalent copper wire reel would reach. The cost seems to be less than that of existing cables. This becomes very attractive to company installations. Some optical fiber strands are just 50 microns in diameter across the core material, have a cladding of a lower index of refraction, and then are covered with a plastic jacket which builds up the size to some 330 microns in diameter (outside measurement). With such a fiber, the light rays must enter the core within the cone angle of some 20° (10° half angle). This gives rise to what is called a numerical aperture (NA) which is the sine of that angle or 0.18. The larger the NA the better the light propagates through the fiber, but this has to have a trade-off. You must balance the losses and distortion that come about with a large NA with the increased efficiency of coupling and the better transmission capability. Proper wavelengths with good windows in the optical fibers being used improve this relationship and make the trade-off less important. The optical fiber light frequencies are in the range from 800 to 1800 nm and some really good windows have been produced at frequencies of 1300 to

1550 nm wavelengths. Losses through these windows are said to be less than 1 dB/km — that is excellent!

The use of optical fibers to connect computers is becoming more and more important. The Japanese (who do much experimentation and have done much development in automation, robotics, and computation) now have a laser printer which can print up to 21,200 lines per minute. Also, they are working hard on the development of optical-fiber cables that can handle infrared frequencies over which computer data can be transmitted from one computer location to another. It is possible that a breakthrough in fiberoptic chemistry might occur which could mean even better light transmission with more windows and less losses in the future.

There is a growing problem in data transmission systems that use light rays: the problem of having light sources turn on and turn off fast enough to handle the bits of data so that they won't be lost, distorted, or mish-mashed together so much that the data cannot be transmitted or recovered. Figure 6-12 shows the relative operational characteristics of a laser light source versus an LED light source.

The LED characteristic curve approaches a somewhat linear span, which means that for a little increase in current applied, there is a very small increase in LED light output. Thus, to get a significant burst of light representing a digital pulse, large change in applied current to the LED must take place.

On the other hand, the laser light source operates on the base of the exponential type curve, near the vertical ascent. Then a small increase in electrical current can trigger a large burst of light energy. That means that in all probability, a faster rate of pulse generation can take place. This assumes, of

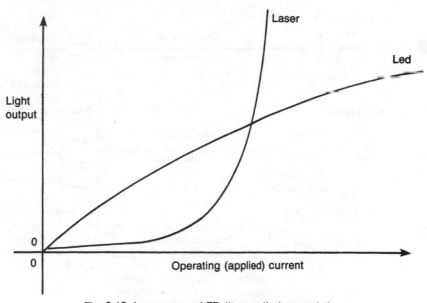

Fig. 6-12. Laser versus LED "turn-on" characteristics.

course, that the light source will turn off as completely and as quickly as the current can be made to decrease. If the applied current is a square wave, then the light burst should also have a square wave type time duration in the optical fiber.

Some systems which have been under development by Bell-Northern Research are shown in Fig. 6-13. At A we have the classic one-way system. The energy goes in on the left and through the fiber to the receiver on the right. At B is a multiplexed system, which may be timesharing or frequency-sharing in operation. As shown, the system would be frequency sharing as the input is modulated with the summation of input signals and on the output the signals are separated by appropriate filters. At C is the two-way system as used by many computer systems.

The fiberoptic connecting link doesn't care which direction the light energy enters or leaves the fiber. It can transmit in each direction if appropriate electronics and connecting interfacing units are used to ensure that it isn't trying to transmit both ways at the same time. Normally a burst of pulse data goes from left to right, then a burst of pulse data comes back from right to left. When this is done at an extremely high rate of speed (with nanosecond timing in the trillionths of seconds) then the system is said to be operating two-way continuously. If mechanical displays and electronic screens display the data, they are actually much slower than the transmission times, so some storage system must be employed to hold the data until it is needed and used.

But if the optical fibers are so small, why not just use two of them on each link, one for transmission each way? That way you'd have the equivalent of separate lines for each direction of transmission and you could actually have

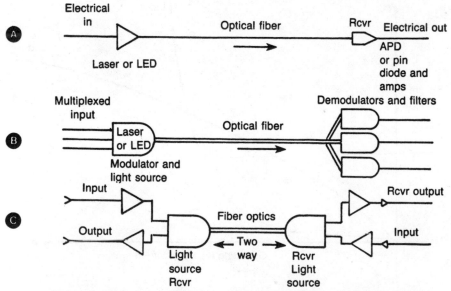

Fig. 6-13. Block diagrams of some fiberoptics systems (courtesy Bell-Northern Co.).

simultaneous transmission going on all the time in both directions. Certainly this is a possibility. Just add a link to Fig. 6-13A, turn it around, put two strands of optical fibers in a cable instead of one, and the job is accomplished. Good thinking.

Now, in order to complete the process, think about reducing the delays that come about in the electronics of the connecting and using units of the transmitted data. There are inherent delays in coils, filters, capacitors, and even in transistors and other solid-state devices. To come near the speed of communication and computation of the human brain all these delays must be reduced and/or eliminated. Isn't it strange to be thinking that the movement of electrons and holes in electric current transmission is slow? But it's a fact, and that is what the future of this science is all about.

COMPUTER PROBLEMS WITH OPTICAL FIBERS

Using optical fibers in computer applications (as of this date) has some problems. One of the major problems is the fracturing of the fibers when they are used in cables and in connectors. This will be overcome by making a better product as time progresses, and by improving the manufacturing techniques as related to computer and automation applications. There are some differences in how the optical fibers are used with communications equipment as required by telephone, video, and long distance data transmission, and the close, short, intricate and delicate applications of a close-in computerized situation. But, research will come up a winner again as it has in the past, given the time to do so, and the money to invest and experiment.

Of course, all the associated equipment (light sources, receivers, and interfacing equipment) that must be used with computers must also develop at a commensurate rate. One thing is certain with all the advantages, operation-wise, cost-wise, efficiency-wise, and time-span-wise, there is no doubt what those micron-diametered optical fibers and all their associated connectors, splices, interfaces, monitoring, checking, and operating equipment will come to be a prime area of consideration within computers now and in the future.

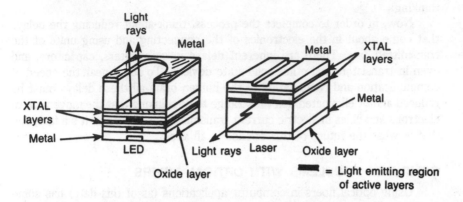

Chapter 7

Light rays

Metal

XTAL layers

Metal

XTAL layers

Metal

LED

Light rays   Laser

Oxide layer

Oxide layer

= Light emitting region of active layers

Cable TV Applications of Lasers and Optical Fibers

T HE RAMAN EFFECT IS A CHARACTERISTIC OF LIGHT FIRST OBSERVED BY C.V. Raman in 1928. In essence, it means the scattering of light from a gas, liquid, or solid with a shift in wavelength in the incident radiation. This is also called the Smekal-Raman effect. Smekal theorized about some vibrational effects of molecules when subjected to incident radiation. The classical theory indicates that the incident radiation, when so scattered, should not be changed. Smekal theorized that the effects of scattering and radiation could be explained by quantum-mechanical theories, and so they were. Certain effects could be observed experimentally (the scattered frequencies related to Stokes and anti-Stokes lines). When the most intense radiations produced by scattering are unchanged, this is known as Rayleigh scattering.

THE RAMAN EFFECT

The day is fast approaching when most homes and businesses will be connected together by fiberoptic cable. When that day arrives, devices will be available that permit interactive communication between homes, libraries, medical centers, and banks. Looking beyond the social, moral, and artistic concepts of this advent in people's lives, you can see a role for lasers and optical fibers. Lasers operate in frequency ranges where extremely wide bandwidths can be accommodated, so multiple signals can easily be handled in some manner. Whether the handling will be in a simultaneous mode or a sequential mode will depend on the type of signals and the number to be accommodated. It is not unusual to place many TV channels on a single optical-fiber cable. The advantage of using such a system to convey these signals is that the system is not subject to electrical interference.

You can imagine the consternation created. Millions of dollars have been spent putting in hard-wire cabling which links homes and offices. One would suppose that companies will not quickly abandon this kind of system to put in the newer and more efficient optical-fiber types of cables . . . or will they? The telephone companies, who pass on such costs to their customers, are doing so, presumably only after much research, cost-effectiveness analyses, and long range operating and usage costs show that the returns are on the positive side of the ledgers. Cable TV may take note of this kind of development.

The demand for cables capable of multi-channel noninterfering signals will continue to grow. With this in mind let's turn to some of the areas of research and development which have been prominent in the field of cable television distribution and operation. These areas point the way to future equipment.

FIBEROPTIC TRUNK FOR CABLE TV

Each fiber of a cable for cable TV carries a single video channel. All the components used were either purchased off-the-shelf or were slightly modified versions of stock hardware. The audio signal modulates an audio subcarrier and the combined video and audio signals modulate the LED source. The bandwidth of the combined signals is five megahertz and the frequency deviation of the FM carrier is plus or minus 12 mHz. At the receiver end of the cable, a silicon avalanche photodiode converts the intensity-modulated optical signal into an FM electronic signal. This is then demodulated in standard form to provide the combined video and audio signals. The systems repeater—needed for any distance of transmission to keep the light levels high enough—consists of a back-to-back combination of one receiver and one transmitter per channel. The optical cable designed for aerial mounting applications was engineered to withstand extreme climatic conditions and it contains a four graded-index fiber interior.

This was developed by Corning Glass with an attenuation of less than 6.5 dB per kilometer. The cable as designed weighs some 64 kilograms (2.2 pounds equals one kilogram) per kilometer and has a tensile strength of 1,000 newtons. Bandwidth of the system cable is in excess of 500 megahertz-kilometers. (Notice how the bandwidth becomes related to distance in this concept.) The cable was strung on poles that were separated some 78 kilometers. Thus it was proved that such a cable was capable of being supported in an aerial fashion and was strong enough not to break or to prevent use in its primary operational mode. Cable General Corporation was responsible for much of this development.

A WIDE-BAND OPTICAL-FIBER DISTRIBUTION SYSTEM

Using optical fibers in a distribution system requires couplings, connectors, and energy propagation. In a study by Technical University of Denmark, the distribution capabilities and concepts, all types of modulation, capacities, and components were analyzed. The modulation characteristics of the solid-state laser were scrutinized. It was shown in the study that this laser is unfit for high-quality analog intensity modulation. Analog transmission can provide only the transmission of a relatively few TV channels at the present required levels of quality, whereas digital types of transmission can accommodate more channels with acceptable quality and a comfortable margin. The study further showed that it was not yet possible to simultaneously transmit many TV channels in such a system and that the replacement of a conventional type of TV distribution system with optical fibers cannot yet be accomplished.

However, in another study and experimental set-up in Ontario, Canada, a TV supertrunk using optical fibers was installed and carried 12 color TV signals or channels and 12 FM channels (stereo) over an 8-fiber cable some 7.8 kilometers in length. The system used for transmission was digital and all signals were digitized for multiplexing into a 322 megabit stream that modulated an injection laser diode transmitter. The National Cable Television Association was responsible for a technical paper presented on this subject. In a multiplexing concept, the necessary signals are transmitted over an optical-fiber cable system. When digital bits are so transmitted, consider that they are serialized or transmitted in a sequential fashion. But, when transmitted fast enough—as these were—then the whole transmission may be considered almost simultaneous as far as equipment and human reception is concerned.

AIR HAZARD PROBLEMS WITH CABLE TV

Another problem which may be solved using optical fibers for cable TV is the problem of radiation on some aircraft frequencies by existing cable TV systems using hard wire, and electromagnetic waves. As previously pointed out, the use of light frequency not only reduces the possibility of interference but also reduces the possibility of radiating electromagnetic signals. This may be another way to counter the aircraft interference problems encountered in existing TV cable installations. In 1979 the FCC formed a committee to study the air navigation hazard problem.

USING A HUB OPTICAL-FIBER CABLE TV SYSTEM

Such a system was proposed and its study indicated that 100 channels could be accommodated using the HUB distribution system and optical fibers. This system presupposed that pulse-width modulation could be used for ease in switching and that a simple set-top converter could be used to deliver pictures to a standard TV receiver. In this study consideration was given to low-cost connectors which were deemed practical for this application. The study was done by Clifford B. Schrock and Associates, Inc.

USE OF CABLE TV FOR SECURITY PURPOSES

When you see a TV camera in a store or bank, you may not realize that this is a cable system. Fiberoptic cabling might be very effectively used in security applications simply because it cannot be jammed by radiation devices. Usually the cable runs are not long, and the reliability and effectiveness of the system must be extremely high. Optical fibers can provide such a system for operating in harsh environments and avoid many of the problems of hard-wire systems. Such systems usually require a motion detector and visual surveillance of areas. The motion detector can alert an operator when he isn't examining a given area at a given time. Other types of sensors can be adapted to work with such a system also.

A LINEAR LASER FOR CABLE TV APPLICATIONS

General Optics Corporation of South Plainview, New Jersey did some research to determine a laser suitable for cable TV applications. They fabricated a laser that delivered 7 mW of power per face, of linear optical power, with respect to the current applied. Their lasers were fabricated from GaAs-GaAlAs wafers grown by the liquid phase epitaxial technique. By carefully controlling the waveguide dimensions with photon implantation, they were able to substantially improve the linearity. The second and third harmonics were less than 55 dB and 65 dB below the fundamental at 70 percent modulation. That old devil signal-to-noise ratio was less than 65 dB. This laser was used to transmit 12 channels of TV signals through a single optical fiber more than a mile in length without any noticeable deterioration of the picture quality.

A LASER BEAM TRANSMISSION USING
A 400 MHz BANDWIDTH ELECTRO-OPTICAL MODULATOR

The Japanese have been responsible for the development of this concept. The optimum configuration of a total internal reflection laser light beam modulator with surface interdigital electrodes has been obtained. The final specifications for this modulator were an effective half-wavelength voltage of 11.5 volts, a capacitance between electrodes of 9.1 picofarads and a modulation bandwidth of 400 MHz into a proper load impedance. A way had to be found to eliminate the harmful resonances due to acoustic waves excited in the electro-optical crystals by the interdigital electrodes, and this was done. Then a very practical system for transmitting TV signals was built and open-air tested using the modulator. Cross modulation and intermodulation levels were acceptable.

THE INTERACTIVE OPTICAL-FIBER CATV SYSTEM

An interactive cable TV system was created in Japan serving some 168 subscribers, having a cable length of some 380 km, and using a 30 channel capacity network. Such a system is thus not only theoretically possible but experimentally practical as well. The results and the equipment used can be found among the IEEE Transactions for 1978.

LINEARITY

Many of these studies describe the importance of linearity. With digital techniques, some nonlinearity may be tolerated because equipment can compensate for this, but, it becomes more difficult to try to compensate for the nonlinearities of light generation from lasers or other suitable optical-fiber light sources. Much work is going on in this field, both in the development of more linear light sources and modulators. Research continues on systems to accommodate nonlinearities. These methods compensate through modified feedbacks and other techniques for the nonlinearities so the output becomes essentially linear over the range of interest.

OPTIMIZED DESIGN STUDY
FOR A CABLE TV OPTICAL-FIBER SYSTEM

A study was conducted by the Harris Corporation of Melbourne, Florida. This study considered the trade-offs for the application of fiberoptics to CATV with the system divided into trunking and local distribution. The performance and cost comparisons for analog-intensity modulation (IM), analog-frequency modulation (FM), and digital pulse-code modulation (PCM) were considered. The performance and cost of multiplexing for all modulation systems were analyzed. The study concludes that the digital system, multiplexed, with fiber-optics is the best for most trunking areas, and analog approaches are recommended for the distribution on a local basis.

CATV IN EDUCATION

An experimental program was considered in West Germany at the University of Bremen. The authors of the study, Haefner, Issing, and Pruess, considered an interactive system of cable TV that could provide two-way communication. They made a proposal to start a pilot program in Germany with 1,000 to 10,000 homes so that reliable data could be collected on the demands of society for such an interactive cable TV program used in an educational mode. This proposal was prepared using international experience in this area with educational TV, the development of the use of computers in an educational sense and the technological constraints in the development of such a system. The authors designed such a system from the technical viewpoint to provide the pilot data.

This type of system, using the latest advances in lasers and optical fibers, should be forthcoming in the near future. Much has to be considered, not only from the technical viewpoint where simplicity and reliability of technical operation must be paramount, but also from the sociological viewpoint, especially what, when, and how the educational information must be prepared and presented. A side issue is the implication of machine teaching versus human teaching in an interactive system. What is the best mix (if a mix is warranted) of the two methods? This may be one of the greatest areas of technological and sociological advancement in the not too distant future.

SOME TECHNICAL DATA ON OPTICAL-FIBER TRANSMISSION LINES

Some measurements of optical-fiber waveguides in lengths over 3,000 feet show a total attenuation as low as 2 dB/km. This has been improved in the latest types of guides. The bandwidth capability of such an optical fiber guide is around 500 MHz. The operating characteristics of these optical-fiber guides are independent of operating frequencies and temperatures. This technology will experience a snowball effect in the following: the strength of the optical fiber guide, the current development of good reliable laser and LED light sources, the ability to join fiber lengths and to multiplex fibers in junction systems so that feeds to many fibers can be accommodated from just one input fiber, and increasing strength of optical fiber cables made in various forms.

THE TELEPHONE COMPANY AND OPTICAL FIBERS

Even as I am writing this book, several runs have been made by telephone companies to test the fiberphone concepts. One run or link is between Pennsylvania and Washington D.C. Another optical-fiber link is being planned that will link Austin and Dallas Texas. There are others either in the planning or implementation stage in the U.S.

This brings up an interesting concept. The telephone linkages already are so vast they link virtually every spot in the world. Future telephone lines will have every type of information you could possibly desire. Just imagine a range of educational programs from kindergarten to college graduate programs, and other special scientific and art programs!

This can have gigantic implications for the cable TV industry. You and I as consumers can only benefit no matter which way the development goes. How nice it would be, for example, to be able to access any scientific library in the world just by pushing some buttons! You'd be able to access any stored information anywhere to satisfy your own particular field of interest.

SEMICONDUCTOR DIODE AND SPECIAL LENS

The high radiance GaAs-GaAlAs LED with a self-aligning spherical lens was developed in order to efficiently couple it to a low-loss optical fiber cable for communications purposes. In the new configuration, the lens is automatically settled into the center of the light-emitting surface in an etched hole containing a clear epoxy resin. When this lens is coupled to a low-loss optical fiber (NA equals 0.14 with a core diameter of 80 micrometers) a coupling efficiency of up to 80 percent is accomplished. The LED emitting area is about 35 micrometers and the lens some 100 micrometers in diameter. It has a refractive index of 2.0. This experiment used a special lens which was very small and became a part of the light emitting diode's surface. (Although with optical fibers, the special rounding of the ends of such fibers can also produce a lens effect. This effect was not used in this case.)

A FIBER-MOUNTED LED WITH AN OPTICAL-FIBER LENS

The Japanese developed a very small fiber-mounted GaAlAs LED to satisfy the requirement for an efficient source of light with some high degree of reliability for an optical fiber communications system. In this case, they used the end effect of a spherical ended fiber. (This means that they rounded the end of the fiber so it produced a lens.) In this way the LED-to-optical-fiber coupling was within the range of usability.

MODERN TECHNOLOGY USES A LASER TO MAKE OPTICAL FIBERS

In one conference on optical fibers, the concept of using a CO^2 laser as a primary hear source to draw the optical fibers was discussed. It was theoretically determined that with proper selection of parameters, a fairly uniform distribution can be maintained around the cylindrical work-stock. Then an experimental system was constructed using a 90 watt CO^2 laser to draw the optical fibers, and proof of this capability was demonstrated. This conference was conducted at the University of Sydney, Australia several years ago.

SAFETY PRECAUTIONS WITH OPTICAL FIBERS

Looking directly into the end of an optical fiber cable connected to a GaAlAs laser or a high radiance LED is very dangerous. A safety zone has been determined in which the minimum distance from the eye to the end of such a cable or fiber has been determined for an acceptable risk. In the case of a cable with just one multimode fiber, the safe distance is at least one meter. This means you should not look into the end of an optical fiber which is excited by a light source at the opposite end, unless you keep your eye at least one meter away from the end of the fiber. Actually, I don't recommend that distance; I recommend you don't look into the end of any excited optical fiber—ever!

With a single mode fiber operation, the critical distance from the eye to the fiber should be at least four meters. Of course, considering the type of fiber, the attenuation of the light, and other factors, you might not be harmed by closer inspection down to a one meter level, so the study found. They also found that looking into the end of a cable was just as bad as looking directly at the end of an optical fiber strand.

With some kinds of light of certain frequencies, there are many photos of experimenters wearing goggles of various types while working around the light beams. If you must be exposed to laser beams or optical fiber beams, then determine what goggles are used, and use them.

LASER TO OPTICAL FIBER COUPLING WITH A CYLINDRICAL LENS

This type of lens is not the type usually discussed. It may have some body to it, and be shaped on each end so it looks like a cylinder when viewed from the side. Several laser-to-optical-fiber coupling methods have been developed and experimentally checked. A coupling idea using a cylindrical lens is quite simple and effective for enhancing coupling efficiency. A mounting and aligning structure for this laser-to-optical-fiber coupling with a cylindrical lens has been proposed. This hybrid structure uses a copper mount on which two grooves are cut for aligning all optical elements without adjustment. Even though the mounting structure was produced by a machining method, the average coupling efficiency of about 50 percent was achieved by using a multimode optical fiber with a half-acceptance-angle of 3 to 4 degrees. These couplers have been successfully mounted in a hermetically-sealed laser package and they seem useful in many practical applications.

SOME MANUFACTURING CONSIDERATIONS FOR OPTICAL FIBERS

An optical fiber waveguide consists of a light-guiding core, optical cladding, and a protective jacket. The core may be dimensioned to a few micrometers to allow only a single, lowest order, mode of propagation. However, a fiber with a much larger core and a multimode waveguide type offers easier fiber-to-fiber splicing and much better light-coupling efficiency. Low-loss fibers, say with attenuations of 2 to 10 dB/km or better, tend to be expensive and have small numerical apertures. Thus light-coupling alignments must be very exact. High-loss fibers of 100 dB/km may be low cost and have high numerical apertures. Given a choice between low-loss and a high-loss type, it may be more cost-effective to choose a high-loss type, according to a study by Jaeger of Galileo Electric-Optical Co. of Sturbridge, Massachusetts. The choice for long runs is obviously a low-loss type, he concludes. However, for shorter runs, the total system-loss can actually be greater if a low-loss fiber is used because of the coupling losses caused by the low numerical aperture of the low-loss fibers.

A COMPOSITE OF OPTICAL-FIBER GUIDE AND STANDARD ANTENNA

Small electromagnetic probes are required in the study of microwave radiation effects on living organisms. To conduct such a study in one case, a miniature isotronic antenna system has an optical fiber waveguide transmission line which was used to couple the data out of the radiation field.

In many microwave systems, the metallic waveguides may terminate in a solid antenna system that is a fiber-optic type of waveguide. These antennas, made from such materials as Teflon and other dielectrics, radiate fine beam patterns, while still being shaped so that they conform to aerodynamic requirements on a military or commercial aircraft.

The reverse may now hold true. Optical-fiber types of waveguides may feed certain types of antennas which do not readily conform to a physical configuration that permits their fabrication out of dielectric materials. Slot antennas may be one type considered. There may be other types that are just as important in their own special applications.

USING OPTICAL DELAY LINES FOR PULSE CODING

At the Illinois Institute of Technology in Chicago, a study was made of the use of optic delay lines to form various kinds of pulses to be used in pulse coding schemes. It was found that optical fiber delay lines offer a simple technique for generating pulse-position codes. It is capable of dividing a single output pulse into a series of equally-spaced pulses that can then be individually modulated with respect to a time base. With a second set of delay lines in a receiver, it was possible to receive the original pulse information while sending a lower peak power level. When appropriate low-loss filters are used, a considerably longer delay time is possible for the pulses. The study concludes that by using a fiber-delay line a number of signal processing functions analagous to those employed in microwave and computer applications should be realized.

It would seem from this study that you could have some optical fiber lines of various lengths all excited at one end simultaneously. The propagation of the pulses through the fibers then proceed at some rate according to the fiber type and length. The output ends, if recombined, then present the initial pulse as a series of pulses, all of which are separated in time according to the time-of-passage through its own optical fiber length. Some delays are used ins ending pulses through optical fiber windings as, for example, in a laser gyroscope, so this concept seems an extension of that proven technique.

TENSILE STRENGTH TESTING FOR OPTICAL FIBER CABLES

At the Stamford Telecommunications Labs, a machine that is able to grip the optical fibers in such a way that no damage occurs to the fibers was developed. The machine subjects the gripped fiber to a tensile strength test. How important this is! You must know how to keep those optical fibers from breaking during installation, environmental changes, and usages. This requires special manufacture of the optical fibers, special cladding materials and methods, and a special jacket. When considering the use of optical fiber cables and trunking circuits, it is important to know what effect various stresses have on the optical fiber section, and how this may effect the transmission efficiency and accuracy of the data transmitted. The Stanford Study presented some results in this direction.

Also, Siemens Research Labs of Munich Germany made many experiments on different glass fibers when they were periodically distorted mechanically. They found a correlation between wavelength and the amount of distortion, and the modes used.

HOW ABOUT A LIQUID-CORE OPTICAL FIBER?

Siemens Labs of Munich, Germany conducted experiments with this concept. They found that whereas straight liquid-core fibers have the capability of transporting linearly polarized light without any incident polarization angle, bent filters exhibit some birefringence with the principle axis fixed to the plane of curvature. The effect of this birefringence can be compensated for by either periodically altering the plane of curvature by 90° or by winding the fiber into a helix. A liquid-core optical fiber has been conceived, used, and subjected to experiments in light transmission — a fascinating concept.

HOW ABOUT TAPPING INTO AN OPTICAL-FIBER WAVEGUIDE?

This idea has been investigated. If you have an optical fiber waveguide transmitting a certain amount of power, it might be useful to tap into that waveguide and extract a given amount of power for some other purpose. A method of tapping by laterally displacing the butt-joined waveguides so that the displacement defines the amount of power transfer that takes place has been defined. In other words, you must misalign the butt-joint by the amount you desire and that governs how much of a light transfer takes place. This is a simple and effective method if you have the means of adjusting the very small fibers or the optical fiber cable. Recall that these fibers are as small as a hair in diameter, and that butt-joining them is a formidable task.

NEAR INFRARED SOURCES IN THE 1 TO 1.3 MICROMETER REGION

Low-loss glass optical fiber waveguides are an excellent media for Raman lasers and amplifiers in the near infrared region, according to a study at the Bell Telephone Labs at Holmdel, New Jersey. Wavelength emissions in the 1 to 1.3 micrometer range is readily available by sufficiently stimulating Raman scattering in single-mode silica optical fibers. After experimenting and observing four orders of Stokes radiations, it was concluded that pulsed, tunable, stimulated Raman emission in this wavelength region is possible using an infrared tunable dye laser near 1 micrometer as a pumping source.

In another study by the Canadian Department of Communications at Ottawa, Canada, the frequency spectrum of a continuous wave fiber-optic Raman laser was measured and observed to consist of several discrete lines covering the 3 nm spectral range. The output of the laser also contained damped transients that were attributed to strong pump-depletion effects in the filter.

Mentioning the frequency spectrum of optical fibers capability brings to mind the following information that might be useful.

Ultraviolet spectrum

- ☐ 200 nanometers and below
- ☐ 200 to 320 nanometers
- ☐ 320 to 400 nanometers
- ☐ 400 to 700 nanometers

Infrared spectrum

- ☐ 0.7 to 1.2 microns
- ☐ 1.2 to 4 microns
- ☐ 4 to 15 microns
- ☐ 15 microns and longer

(The divisions are shown in the ranges that many commercial companies use; much of the technical literature available considers these divisions for specific applications.)

A FIBEROPTIC DIGITAL DATA AND TV LINK

In communications techniques there is a concept called data compression, which means squeezing information into a smaller frequency range than it normally occupies. In the concept developed by Bell Telephone, this technique will be used to squeeze nine bits of information into an eight-bit code (the maximum code-word length that an optical fiber can handle reasonably well and maintain the quality required for the transmission of the data). If a nine-bit analog-to-digital data link were used, the optical fiber link could not handle the high-speed data rate involved without other problems developing.

A LASER READER FOR HOME RECORDERS

Actually a laser doesn't read information. It supplies the light source and then optics, optical fibers, and photodiodes help convert the reflected light pulses back into digital signals. These are converted into analog signals by digital-to-analog techniques, amplified and presented to display or audible output devices.

Such a digital system is being developed in Japan by Matsushita and will have a stereo signal consisting of a 16-bit code sampled at a 44.056 kHz rate on 12 tracks on each side of a cassette for a total of 24 tracks. The prototype system which Matsushita has produced uses micro-cassettes. It seems to work fine except that it uses a slightly smaller code. It uses 14 instead of 16 bits and the sampling is at a reduced rate of some 33 kHz. The dynamic range is almost equal to 85 kHz. Lasers used in this application must be small, reliable, and arranged so that they can track the microgroves that provide the reflective pulses needed to complete the system's operation. In the first stages of such a system, where high sensitivity is required, that the use of optical fibers (immune to electrical noises) might be used.

TESTING OPTICAL FIBERS

How do you test an optical fiber or optical-fiber cable? You need to know the attenuation per kilometer, and the purity and frequency of the source.

An instrument for the nondestructive measurement of attenuation in optical fibers was developed by the Stard Telecommunications Labs, Ltd., of Harlow, Essex, England for checking the system laser and the light detector equipment. This instrument was the first of a series designed to cover all tests required for optical fibers. It incorporated its own gallium-aluminum-arsenide (GaAlAs) laser source. The attenuation of optical fibers with some range of core sizes, and whose losses might range as high as 60 dB, may be measured within 1 dB accurac with the instrument. The method used is electrical substitution. A comparison is made with a short (zero dB loss) section of similar fiber. Then the system detector can be checked for a go no-go kind of performance. Also, the detector can be calibrated to check the system's laser source.

In all of the new installations of optical fibers, the method of testing and checking for the three very important system elements (the light source, the optical fiber cable or strand, and the detection system) will be mandatory. Not only will the light source have to be tested for power output capability, but also for generation of the proper light frequencies, polarities (probably), monochromaticity and mode. Instrumentation for detection of faults, flaws, and location of abberations will be required and designed, expert, intelligent personnel will be required to operate such instruments.

A study was made by the Italians that involved the determination of the error probability in digital transmission over optical fibers. The method took into account the principal phenomena that influences the error probabilities. First an approximate expression is obtained for the error probability, based on simplifying assumptions and hypothesis. Then the exact procedure and approximate expression are applied to a particular case. This case assumes that the amplifier-equalizer has a transfer function derived form the hypothesis that the isolated impulses both at the output and at the input of the receiver should have a Gaussian shape.

It is easy to see that all theoretical and practical methods available are being used to determine the capability of the optical fiber transmission systems concepts. When finally you have all the equations and instrumentation required to design and test such equipment, then optical-fiber system design will take its place alongside microwave systems and hard-wire systems.

GRADED INDEX FIBERS

The requirement in many applications is to have optical fibers with a graded index of refraction. Let's consider how to make a fiber with different gradations of the index of refraction outward from the center of the fiber core. It has been possible to make such a fiber by using a rod of a lower index of refraction and then cause it to shrink onto a rod of a higher index of refraction. Thus the index of refraction gets smaller in a step-wise fashion out from the core center towards the surface.

In the old days of guided missiles, a system was proposed using a Luneberg lens as a focusing element for radar waves both in emmision and returning echoes. The idea in this lens was that it would be round (like a ball) and have a variable index of refraction for microwaves. The index of refraction was changed by the use of some rods of various sizes in the ball's construction. Also, it turns out that no matter from what direction the microwaves entered the lens, they would come to a focus at a point (or small area) directly opposite the direction the rays entered the ball. (That was why it was called a lens.)

Optical fibers can be made so that the index of refraction varies either in a gradual manner or in a step manner. Bandwidths of graded fibers are very much larger than can be obtained with step-type fibers, but the numerical aperture for a graded fiber is less than that for a step-type fiber. This means a more accurate positioning of the light source (to get the light rays into the fiber) is required with the graded fiber types. The attenuation of the light ray sin a graded fiber is less than it is for a step-index type.

SOME DATA ON AN EXPERIMENTAL
OPTICAL-FIBER SYSTEM INTEROFFICE TRUNK

The basic elements of such a system are a light source, a means for modulating that light source, a transmission system (optical fibers), and a receiver to convert the light pulses (assuming digital transmission which is most predominant currently) into electrical signals, and the shaping and reconversion of the electrical signals into some usable form. In this experimental operation the laser light source used was a GaAlAs injection laser. Avalanche photodiodes were used as receivers. The light source from the laser was electronically monitored, controlled, and modulated. The gain of the receiver was controlled by varying its bias supply. An avalanche photodiode in a typical circuit may use a bias supply that can be so set that it gives a threshold value of detection for incoming light pulses. A high bias would cause an intense signal to be present before it was recognized and a threshold value of bias would result in the utmost sensitivity for the detector. You might want to file away this information if you experiment with lasers, photodiodes, and optical fibers.

SOME PRINCIPLES OF OPTICAL-FIBER CABLE DESIGN

The problem of making optical fiber cables was studied by the Stan Telecom Labs Ltd. at Essex, England, and they came up with some design criteria which seems appropriate. In their study the main concerns were brittleness of the optical fiber material, optical sensitivity, and the mechanics of manufacture, installation, and servicing of such cables. The necessary mechanical strengths and properties for the cabling of the optical fiber light-guides was obtained by the application of plastic coatings. A number of such coated fibers may then be combined with a base cable stringer or strengthener that increases the total strength of the assemblage. This is done with such fillers and wrappings as required to maintain the geometric register. The unit may then be plastic wrapped and it can be covered with water barriers and given even more strengthening members so a large tensile strength can be obtained. Many such optical-fiber sections can then be combined into complete cables that have a very high capacity for information transmission. The basic design requires that the fibers be protected against tensile and radial stresses, bending conditions, and environmental problems. If you have ever examined the optical fibers used in light displays at novelty stores and have taken one such fiber out of the display to look at it more closely, you'll find that it is very small and very brittle—it breaks easily. Of course, these aren't the fibers used in cables, but this does give us some idea of the problems associated with making a good strong optical-fiber cable that can be handled in the field like any other cable.

LEAKY FIBERS

There is such a thing as a leaky fiber. The loss of the light rays of some given frequencies constitutes the leak, because these rays propagate at some point at more than the critical Brewsters angle and so escape into the cladding and are absorbed or attenuated very drastically.

There are leaky microwave systems, some of which are designed intentionally to leak radiation. There are situations where the microwave energy gets out of a waveguide at a joint or a twist or some other discontinuity in the waveguide structure. Light rays can be treated as just smaller wavelength electromagnetic radiations. Thus, anything that would provide any kind of discontinuity or imperfection in the fabrication of an optical fiber might cause the optical fiber to leak energy at that point. Depending on the frequency and mode(s) of operation of the optical-fiber guide, some frequencies may leak and some may not. In some cases it has been found that undesirable radiations can be attenuated by this leaky process. This is good, because it reinforces the desired radiations.

THE LIFESPAN OF INJECTION LASERS AND LEDS

Some studies and experiments at the Princeton, New Jersey RCA Labs indicate that some lasers (GaAlAs types) may have a very long lifetime for useful outputs. In some tests these lasers have been operated for longer than 10,000 hours and the study, which was conducted some time ago, indicates that LEDs and cw lasers may have a useful lifespan of over 100,000 hours. This becomes very important when considering small lasers, and LEDs being used as an integral part of an optical-fiber cable transmission system.

SERIAL-ACCESS COUPLERS AND DUPLEXES FOR OPTICAL FIBERS

These can be obtained for single-strand multi-mode types of optical wave-guides by using conventional optical components such as beam splitters and lenses. This approach is difficult because of the small optical fiber diameters and the extremely close tolerances required during fabrication and assembly. Some alternatives are the taper fiber coupling and the welded access (T) coupler. In single-strand fully duplex systems, the amount of light reversed or backscattered as the optical pulse travels down the fiber is an important consideration. The tapered fiber coupler has been demonstrated as a convenient means to experimentally measure this backscattering effect.

A RADIATION-HARDENED OPTICAL-FIBER SYSTEM

The Harry Diamond Labs at Adelphi, Maryland developed a radiation-hardened optical fiber transmission system with a 400 MHz bandwidth and linear response. An amplitude modulated cw laser diode is used in the transmitter with automatic bias compensation applied instead of temperature control. This system can select, via an optical-fiber control link, the attenuation between the transmitter and the receiver. The transmitter and optical-fiber portions of the system were hardened to withstand a photon irradiation of more than 5×10^5 rads from a Betatron.

COUPLING A LASER INTO AN OPTICAL FIBER

Two methods of coupling a multimode fiber to a GaAs double heterostructure laser were investigated; first, direct coupling into a plane-fiber-end-face, and second, coupling into a cylindrical glass-fiber lens. The power coupled was measured for three different lasers and one laser was adjusted by its bias current. Some optical feedback was observed when the coupling between the laser and the optical fiber end varied just a few tenths of a micron. The tolerances for the fiber end type of coupling are, as expected, very tight.

MULTIMODE AND SINGLE MODE

The optical fiber may have only one frequency propagated through it (singlemode) or it may have many different frequencies propagated through it (multimode). If a pulse is propagated in multimode form, some frequencies may propagate faster than others, so there is a widening of the pulse. The refractive index governs, to a large extent, this effect. Stepped or graded optical fibers may help to avoid this problem.

THINKING IT OVER

A new era in communications and computer techniques is upon us. In years past it was predicted that we would someday have solid blocks from which we could get sound and music, and by looking at the block we could see various images. It seems that such a day is at hand. The solid-state devices considered in this text approach that solid block condition of electronics.

No longer will we be concerned with capacitors, resistors, wiring, and soldering. Everything electronic will one day become a chemically and atomicly bonded solid and do miraculous things for us. Computers will shrink to the size of a playing die and be able to access other computers and memory banks everywhere through means we cannot now envision. Who knows, perhaps someday in the future we will be able to learn in one year what we learn in 20 years today! How strange it will seem to wake up some morning, after setting the educational knowledge brainwave organizer the night before, and find oneself a dentist, doctor, engineer, physicist, aeronautical engineer, or whatever! Those little die cubes will have accessed stores of knowledge everywhere and caused that information to be implanted in our own changed brain contours and patterns.

Lasers are rapidly becoming standard equipment for most audiophiles as the compact disc player and laser discs become more readily available and affordable. The future holds much promise for the application of the laser in many services of mankind. As this instrument develops from the laboratory stage to the home owner, most of us will need to be educated to the practicle side of the laser. One of the most promising uses is currently for communications over fiberoptic cable as can be witnessed by the advent of new long distance telephone companies relying totally on this current technology.

The miracle of optical fibers is permiting more information, intelligence, and data to be transferred from one point to another more quickly and precisely than ever thought possible. As their algorithms and equations are developed and proved, fiberoptics will become as wires, easy to construct to precise tolerances, accurate and indefatigable in operation, and a type of circuit which will enhance our lives and ways of living.

It is good to know about and understand fiberoptics and lasers even though you may not be able to experiment with them. You may not be able to

182

get the fiberoptic material and all the necessary experimental equipment. But time will solve this latter inconvenience. Strands of optical fibers used in various capacities (from kits for the younger experimenters, to various devices using optical fibers for the more advanced) will become very common in the future. More and more uses for these solid-state devices are becoming available.

Chapter 8

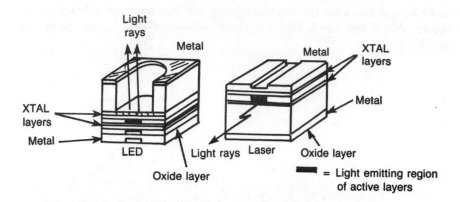

Light rays

Metal

Metal

XTAL layers

XTAL layers

Metal

Metal

LED

Light rays

Laser

Oxide layer

Oxide layer

= Light emitting region of active layers

Experimental Applications of Fiberoptics

BEFORE GETTING TOO DEEPLY INVOLVED IN THE THEORETICAL ASPECTS OF light transmission, light-guides, light generators, receivers, and such, let's take a breather for a moment and consider some uses and applications for light beams and light rays with which you might experiment. I have not conducted the experiments, nor fabricated the equipment described herein. But most of what I describe has been built and fabricated and operated by firsthand investigators. A word of caution: If a device doesn't do exactly what you expect it to do when you build it, do not hesitate to change and modify it as necessary to make the end justify the means! That is more than half the fun of experimentation. Keep a notebook. If you choose, let me hear from you about your efforts.

A LIGHT-FOLLOWING ROBOT

Karol and Mila Nowakowski have a robot with a phototransistor in its base. When the robot is in motion, the phototransistor senses an illuminated line on the floor. The base of the robot has a black felt skirt to conceal the sensors and other equipment, and to eliminate ambient light from the robot's surroundings. (See Fig. 8-1).

The photo-transistor is set into a section of black (probably rubber) tubing so that it has a narrow field of vision, as shown in Fig. 8-2, a typical circuit that produces a current output when it has a light input. If you are going to steer a robot in this manner you actually need two such sensors. One sensor must be positioned so that if the base goes to the right, the electrical output of the

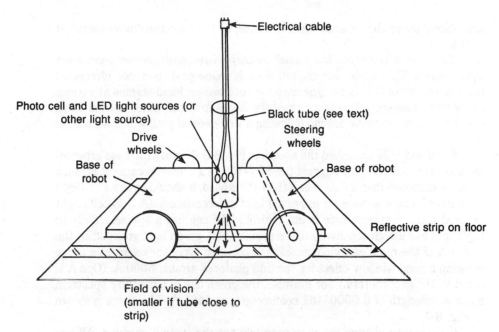

Fig. 8-1. Robot's base.

photodiode will (through suitable amplifiers) be capable of operating a control circuit to cause the steering motor to turn the base so it moves to the left, or back on track. The same effect (in reverse direction) would apply to the second photodiode sensor. That way steering is accomplished. When both photodiodes get the same amount of reflected light or no light, the base steering control

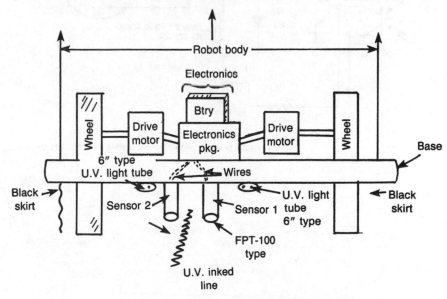

Fig. 8-2. The Nowakowski line-following robot.

unit should make the steering wheels center, so the base thus moves straight ahead.

The base of the robot has a small battery-powered ultraviloet fluorescent light source. The guide line on the floor is made with invisible ultraviolet fluorescent ink of the same type used for readmission hand-stamps at various social and commercial functions; thus, the line is invisible under normal viewing. The robot can move around following a well-defined path which we cannot see.

Karol and Mila furnished the sketch in Fig. 8-3 showing the base arrangement of their robot. They suggest the FPT-100 as a sensor because it is much faster in response than a CdS cell. If the FPT is used, it should have a U.V. light filter over its face section for improved light differentiation. (A CdS cell might be used if the response time is not critical.) The cell has a low sensitivity to U.V. light and a high sensitivity to the primary blue-green light given off by the U.V. ink of the most easily obtainable and common type, they sent us a curve showing a good window effect for the CdS photocell around 5000 Å. (One Å is equal to 10^8 centimeters.) For example, the green line in a mercury spectrum has a wavelength of 0.00005461 centimeters, or 5461 Å. The curve is shown in Fig. 8-4.

It is easy to change the path once you get the system working. All you

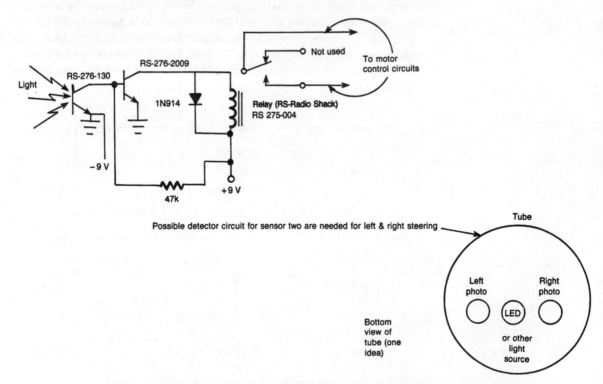

Fig. 8-3. A light sensing concept for a path-following robot.

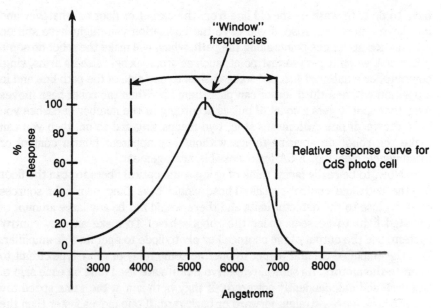

Fig. 8-4. A typical response curve for a CdS photocell.

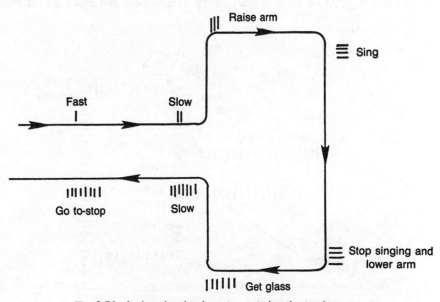

Fig. 8-5A. A plan showing how to control path steering.

have to do is to wash up the old line from the carpet or floor or whatever and ink down a new line. Also, if you have the inclination you might have station information at various points along the path, which will make the robot do some other task when it gets to that point, such as stop, speak, raise its arms, sing, compute, or whatever! Just add some dashes to one side of the path line and in a position where a third sensor can pick them up. When the robot base moves over that spot it gets a count of pulses according to the number of dashes you have drawn. If one dash means stop, two means sing and so on, then you can make the robot do amazing things without any apparent human control or intelligence. See Fig. 8-5A for a possible arrangement.

Now, to be really fancy, think of using some optical fibers to scan the floor for the path and control signals. These would have their own light sources built-in close to the detector units and there would not be any large amount of radiated light to be seen under the robot's base. To make an easy control system, use the output of the photocell or photodiode to operate a dc amplifier. Use the amplified current output to operate some relay or relay-type circuit to operate the motors. As shown, two drive motors are used (one on each side of the base and independently powered). If they both run at the same speed the robot's base moves straight forward or backward. If one moves faster than the other, the base turns toward the slower wheel.

You might make a path of spaced pulses and use a pulsed motor or motors so that the synchronizing problem is not so great. (See Fig. 8-5B.)

Each time the base passes over a pulse both motors advance the same amount except on turns, when the side control signal can insert some resistance onto the lead of one motor so it turns more slowly and causes the steering required for turns. Yes, it will take some experimentation. The ideas are all sound and it could be a fun project to try to assemble and make operational.

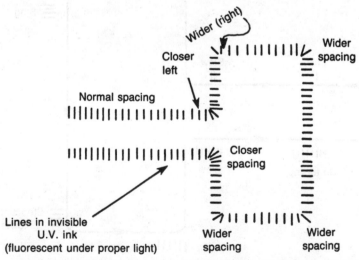

Fig. 8-5B. Making a path that can generate electronic steering pulses.

Also, there is plenty of learning and education to be derived from the effort. Good luck!

SOME SUGGESTED CIRCUITRY FOR A PATH-FOLLOWING ROBOT

At the outset let me call your attention to two books on robotics written by Edward L. Safford, Jr., and published by TAB Books Inc., Blue Ridge Summit, PA 17294. They are *The Complete Handbook of Robotics* (#1071), and *Handbook of Advanced Robotics* (#1421). These could be of help to you if you investigate the robotics projects. Also I recommend that you obtain some literature on all possible electronics circuits that show comparison amplifiers, digital to analog circuits, analog to digital circuits, and control circuits for steering and motor control such as those described in TAB books numbers 1174 and 1135.

Before examining some circuitry, let us also consider another path arrangement as shown in Fig. 8-5B. This diagram shows how you might draw short cross-wise lines with your invisible ink that are fluorescent and can be seen by the photodetector. What happens is that as the robot passes over the line it sees a series of pulses. When the pulse strength and rate is equal into both photodetectors, then the robot base goes straight. If you close-space some at a turning point, then a counting circuit inside or in series with the photo output causes a turn-over-ride signal to be generated so the base turns in a given direction so long as that fast pulse rate exists. A return to the normal rate (governed by how closely or far apart you space the lines) again releases the steering control circuits to make the base go straight. To make the robot

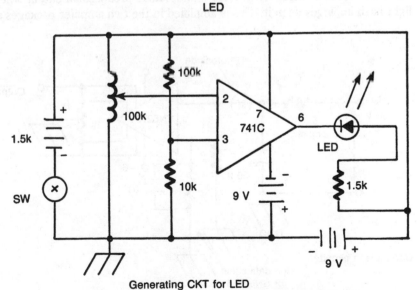

Generating CKT for LED

Fig. 8-6. A light-generating circuit used with an LED.

base turn in the opposite direction, you need another type of signal. This can be accomplished by using a wider spacing (slower pulse rate pick-up). That signal could mean go in the opposite direction until the normal rate is again seen by the robot.

It can be very interesting to imagine what you can do with variations of the pulse rate, from wide (long duration) pulses, slow pulses, narrow pulses, and fast pulse, to a special coding of pulses as laid out on the floor or path. With a proper circuitry and proper coding you might be able to pre-set many commands for your robot. Once activated, the robot starts moving along a given path stopping at various places to do a multitude of things such as singing, or opening doors. The capabilities are only limited by your imagination and electronic genius. Use a path like that shown in Fig. 8-5B and anything is possible!

To get you started on such a project I've included some circuits in Figs. 8-6 through 8-9. The three circuits shown in Figs. 8-6, 8-7, and 8-8 can be used in various ways. The LED circuit (Fig. 8-6) can be used to cause an LED at the end of a small tube (isolated from a photodetector) to illuminate the floor when placed so it is in close proximity to the lines. It has a sharp and small field of vision.

The circuit shown in Fig. 8-7 can be used if a balanced output level is needed and this level can be affected by the light falling on the photo-detector. If the light is more intense, a larger output of some polarity might be obtained. If the light is less than a given level, the output might change polarity and be smaller. You have to experiment with these circuits to see what they do and what you can do with them.

The circuit shown in Fig. 8-8 can be used with the circuit in Fig. 8-9 to cause the robot to steer itself. Here a photodiode produces an output when a light flash impinges upon it. This is amplified in the two amplifer packages and

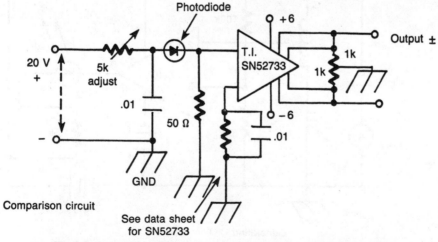

Fig. 8-7. An analog comparison output circuit with a light-detector input.

190

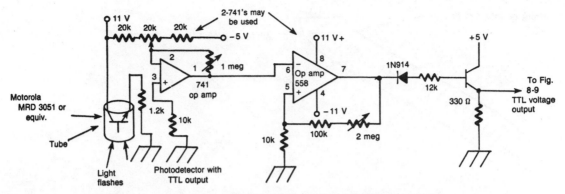

Fig. 8-8. A photodetector-to-TTL type output.

sent to a transistor output that can be coupled to the input circuits of Fig. 8-10. Two such circuits are needed for steering, as you need two photodetectors. If you use a placement of side pulses as mentioned earlier, then you might use a third circuit like this with a photodetector positioned to see those pulse marks. Some component values are shown and these might be okay, but don't hesitate to change them or experiment to get what you want—and that is a voltage or current pulse at the output for each line segment seen by the detector photodiode.

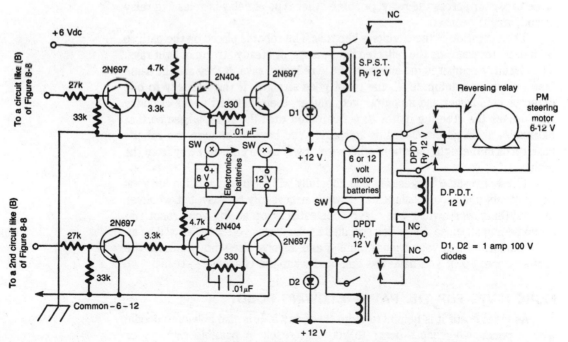

Fig. 8-9. A possible steering circuit for a path-following robot.

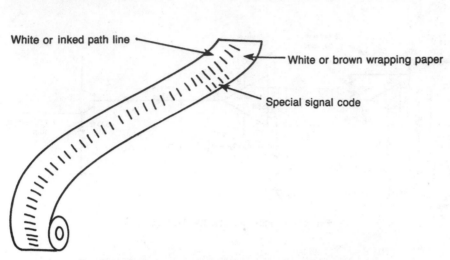

White or inked path line

White or brown wrapping paper

Special signal code

Fig. 8-10. Using a roll of wrapping paper for the robotic path.

Finally, some kind of steering circuit is needed. I show a possible one using 12-volt relays which can be transistor activated. I show some types of transistors but others of an equivalent type certainly can be used also. The idea is to take the output from Fig. 8-8 and feed it into this circuit in such a manner that an appropriate relay closes and holds closed as long as the input continues at a given level. I do not show any holding or delay-type relay arrangement. You may need to modify the circuit to get this to hold over the pulse spacing. A large capacitor across the relay, or some other type of delay-in-releasing relay circuit, might be used.

The operation of the circuits is like this: The robot is placed on the path so both detectors can see the reflected light pulses, or steady stream of light rays. The circuitry compares the intensity of the reflected rays. If they are the same (and below a maximum level) the robot goes straight. If the intensity to one detector gets larger and its output gets higher, then that part of the circuitry that causes the steering motor to activate and turn the robot's base so that balance is again accomplished, is activated. You have to be sure the robot's base turns in the right direction; otherwise, it will move entirely away from the path!

These are some suggestions and hopefully will form a foundation for your experiments and accomplishments. It won't necessarily be easy, but whoever said anything worthwhile was easy? Your satisfaction seeing your robot run around doing all those things you specified at the places you specified will be an utter amazement to your friends, family, and anyone else who witnesses the performance. So, if you have the time and want the challenge—try it!

MORE HINTS FOR THE PATH-FOLLOWING ROBOT

As ideas occur it is helpful to write them down. It is also helpful to doodle with a pencil when ideas occur to you that result in possible circuitry or construction of a path-following robot. One such idea just blossomed in my

192

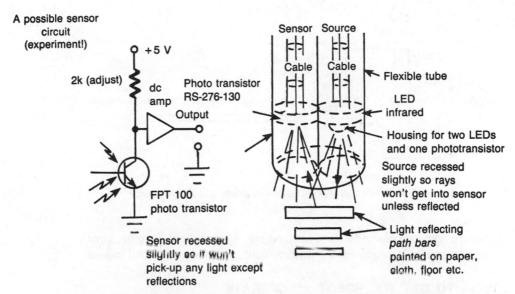

A possible sensor circuit (experiment!)

+5 V

2k (adjust)

dc amp

Photo transistor RS-276-130

Output

FPT 100 photo transistor

Sensor recessed slightly so it won't pick-up any light except reflections

Sensor   Source

Cable   Cable

Flexible tube

LED infrared

Housing for two LEDs and one phototransistor

Source recessed slightly so rays won't get into sensor unless reflected

Light reflecting *path bars* painted on paper, cloth, floor etc.

Fig. 8-11. Drawing of a possible phototransistor-LED path probe for a path-following robot.

mind. It would be nice to have a path for testing purposes which wouldn't require painting white strips on the floor, or putting invisible ink strips on a rug or such. So here is this idea for what it is worth to you.

Purchase a long roll of wrapping paper that is a dark brown and wide enough to form a path when properly marked. Being paper, it will be easy to paint white strips on it or put any other color marking which you might want to try for a robot sensor to detect and follow. When you want to test your robot machinery to see if it will actually come anywhere near doing what you want it to do, stretch out the path segment (any length you choose), place the robot on the paper with its marked path and turn it on to see what happens. Your family will love this idea because it doesn't mess up home arrangements. Simply roll up the paper when you've finished that experiment and you can use it again as needed later on. Also, it works well inside or outside the house, on a porch, in a garage, on a sidewalk or wherever. Figure 8-10 illustrates this idea.

SENSING LIGHT REFLECTIONS

In your experimentation you may find that fluorescent lighting may not work well. Instead you may want to develop a special light-emitting probe with a closely built-in sensor of the phototransistor type. One such type of phototransistor is the FPT 100 or Fairchild FTK0031. These phototransistors work best with a standard incandescent light or with an infrared-emitting LED. Probably these phototransistors will work all right with something like an automobile light that operates from a battery—it is worth trying. In any event, you might make a probe such as illustrated in Fig. 8-11 and see if it won't pick up the light flashes or reflections from the path marking well enough

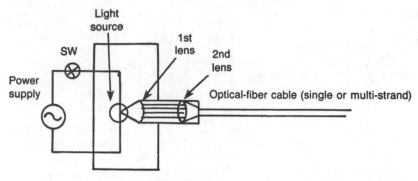

Fig. 8-12. Illuminating an optical fiber with lenses and a light source.

to produce some signals to be used for steering. As you might imagine, a path using this type source and sensor arrangement could be very small indeed.

TRYING TO GET THE ROBOT TO OPERATE

Once you've got the equipment so that pulling the paper path under the sensors causes the steering mechanism to move the wheels fast and slow or turn them, then it is time for a trial run. Adjust the path for a straight run first, just see if the robot will follow a straight line and be able to steer itself to stay on that straight line or path. Then, if that turns out okay, try a gentle turn and add some extra control signals. Develop the system one step at a time until it all works. By the way, the nice thing about using a sensor that operates on infrared wavelengths is that the sensor can be adjusted so it is not sensitive to ambient light. You might be able to demonstrate the system either inside or outside in broad daylight!

If you are really experimentally inclined, you might use fiberoptics to bring the light down to the path and the reflections back up to the sensor that can be placed at the best possible location in the electronics package. In many applications of optical fibers, multistrands all packaged together are used to form a bundle. This makes it redundant and more reliable. When one strand fractures there are many others to transmit light rays over the cable.

When handling optical fibers which are so terribly small in diameter, it is nice to have enough of them grouped together so that there is something substantial to grasp. A cabled-packaging of fibers might be made up to a diameter of some $\frac{1}{8}$ inch or perhaps under some circumstances to $\frac{1}{4}$ inch in diameter. These are all independent but they also work together to transmit light rays from one point through themselves to another point.

Not the least important of the various items needed for such light transmission are the connectors. These connect the ends of the optical fibers so that the light rays can be brought together and passed into the device that receives the rays. Contact Oriel Corporation, 15 Market Street, Stamford, Connecticut 06902 for their catalog and information on any optical needs you might have. Figure 8-12 shows a sketch of how an xenon or a mercury vapor arc lamp can

194

be coupled into an optical fiber transmission line using some of the special lenses available from Oriel.

There are many types of light sources that produce various frequencies of visible light. Determine the type of light desired (infrared, ultraviolet, or whatever) and then try to get a light source rich in that frequency content. You also need to get an optical fiber that will pass in a mono-mode (hopefully) the frequency being generated. All this takes some consideration, investigation, and planning, but the results are worthwhile.

SOME VARIOUS OPTO-ELECTRONIC CIRCUITRY

Since we have turned to somewhat of a practical approach in this chapter, let us continue the investigation by looking at some electronic circuits that use optical-coupling circuits. One of the first types of circuits (and an effective one) uses a photoresistor element. One type is the cadmium sulfide photocell (Radio Shack 276-116 or equivalent). This cell changes resistivity when light falls upon it. It can be used in almost any circuit using a resistor arrangement as input and gives an output proportional to the light intensity falling upon it. It works with a wide range of light types, from flashlight to ac bulbs. When used in the circuit shown in Fig. 8-13, it can control other devices.

The circuit is a variation of a Wheatstone bridge. This resistance cell is used as one leg of the bridge. The one megohm balance resistor does just that — it balances the output to zero by balancing the bridge input section. The gain control governs the current flowing through the output terminals as the input becomes unbalanced. Thus a transistor connected properly across the

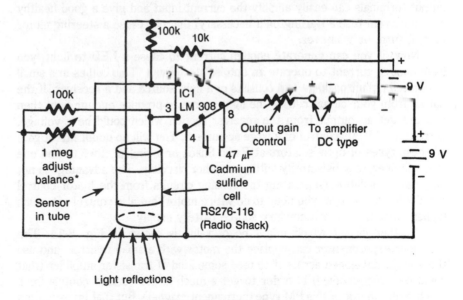

Light reflections

* balance is zero output to amplifier or close to it

Fig. 8-13. A light-sensing circuit using a cadmium-sulfide cell.

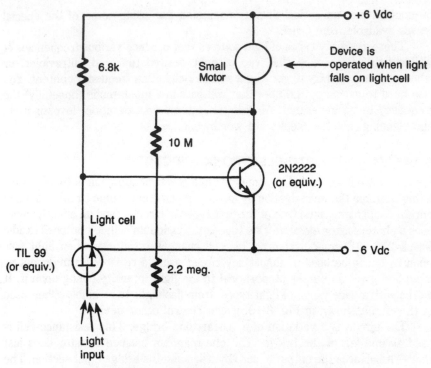

Fig. 8-14. A simple light activated control circuit.

output terminals can easily amplify the current input and give a good healthy output current (use a Darlington if necessary) that can cause a steering motor to run, turn, or whatever.

Now, if you can generate enough current to cause a LED to light, you have enough current to operate an optoisolator circuit. This comes in a small integrated circuit package and consists of a light source and a receiver. If the output from your circuit causes the light source to produce enough light then you can get an output from the receiver section which could be a voltage, current, or TRIAC control of a large ac current. You will no doubt investigate all these types of opto-isolators at Radio Shack or a similar store to find one that will operate satisfactorily with your other circuitry. The advantage is not only circuit isolation (separating the sensing circuits from the motor control circuits to do whatever you need to do with a motor) but also control of a much higher current from a secondary source battery supply.

Another circuit which might be of value is shown in Fig. 8-14. The ingenious experimenter can replace the motor with a small resistor, and use the voltage developed across it to feed some kind of transistor amplifier (that could be direct-coupled) in order to get a much larger current control for a much larger motor of this PM type (permanent magnet). But that investigation I leave up to you. One suggestion, however: if nothing else, a small relay with a low resistance coil can be substituted for the motor shown, and that relay through its contacts can able to control a much larger motor's current. Have a go at it!

THE SCIENTIFIC METHOD OF EXPERIMENTATION

Among the gentry there is always much hinted and stated with regard to the scientific method. This is a step-by-step procedure recommended for experimenters and those investigators whose skill level and knowledge level places them a little beyond the ordinary experimenter. (In other words, they seem to have some firm idea of what they are after and they devise experiments, somewhat skillfully (sometimes) to prove or disprove certain ideas.) So here is the scientific method for what it may be worth to you. It is: Conceive the idea or theory. Build the equipment to conduct the experiment to prove or disprove the theory (this may take some doing and some delving into design concepts, and even development of new types of devices and machines to be able to conduct the experiments). Relate the theory to the experimental results by conducting the experiments. Then publish the results in some prestigious scientific journal.

What is not stated in the scientific method process is that one cannot prove something is absolutely true. To prove something is false requires one and only one example. Truth in science is philosophically impossible to show. You may have to conduct one million and one experiments in order to show possible truth. If the hypothesis is possibly true, then you has to come up with equations that permit others to duplicate the results obtained. Also, you probably have to be prepared to defend your work against the innumerable attacks that always come upon your experimental procedures. Are they complete enough? Are they valid enough and so on? You may never know!

OBTAINING SOME OPTICAL FIBER FOR EXPERIMENTATION

The next question is: "How can I get some optical fibers to do some experimentation with?" The answer, is to write to Edmund Scientific Com-

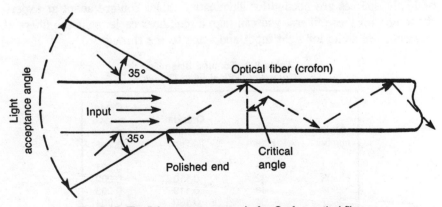

Fig. 8-15. The light acceptance angle for Crofon optical-fiber.

pany, 7785 Edscorp Building, Barrington, New Jersey 08007, and ask for their catalog in which optical fibers are listed and information on optical fiber experiments that you might perform (also fiberoptics kit #70,855) is provided. That is a start. Then, for lenses and light sources, contact the Oriel Corporation, 15 Market Street, Stamford, Conneticult 06902. This may give you the materials you need to begin your investigations.

Edmund Scientific has some light fibers bundled into cables. From 16 to 64 of such fibers are sheathed in a plastic covering and can be used in experimentation. What experiments? Well, vision — that is, looking into one end and seeing what is near the other end for one thing. You can bend these fiber cables around and still easily see what the other end sees within its acceptance cone of some 70°. You might look into walls, into machinery, into people, or whatever and see what is going on in there. You cannot do this normally because you have no way to get the reflected light from the inside of things to your eyes, which are on the outside. So, you can use one cable to send light rays through into the darkness and use a second cable to look through to see what is there.

What a nice idea. In some homes you might want to look at electrical wiring, or plumbing, behind the sheetrock walls or see if a rat's nest is located there. You don't want to tear out the wall, or cut a big, unsightly hole in the wall to look around, so use optical fiber cables. These take only a very small hole to get the light source cable and the seeing cable inside to look around. These holes are easily patched if nothing is there. If something is there that needs correction in some way, then you can open the wall up for access knowing that you aren't just tearing out some wall needlessly.

Edmund Scientific calls their light fibers Crofon which transmits light uniformly in the visible region, reduces transmission sharply in the ultraviolet region and absorbs most of the light in the infrared region. The material is relatively strong and reasonably larger in size (a single fiber being some 10 mills in diameter). To give you some idea as to the relative size of the cables they produce see Table 8-1.

Fittings for an eyepiece are required for the cable used to see through, and light sources are needed for illuminating cable. You are adept at experimenting with these fibers, you can take a multifiber cable, split the fibers at one end, use some for light input and some to see through.

Table 8-1. Edmund Scientific Company Optical Fiber Cables

Stock number	Fibers	Outside diameter (inches)	Cross section (square inches)
2504	16	0.087	0.001
2505	32	0.110	0.002
2506	48	0.119	0.003
2507	64	0.130	0.004

USING THE OPTICAL FIBERS IN EXPERIMENTS

How do you handle an optical fiber cable? You first cut it to the required length using a sharp knife or a pressure cutter that trims paper (a sharp razor blade has been suggested). In any event you will want to get the cable cut to the right length for your use. Second, you need to polish the ends so that the best light transmission is obtained. You can do this with a special lacquer or by some abrasive polishing method. Using some kind of abrasive (such as wet#600 alundum), and carefully holding the end of the cable, cut perpendicularly to the cable length, against a very fine grinding wheel, polish with the required motion. You might polish by wet sanding with 400 and 600 grit abrasive paper and then finish the polishing by cleaning thoroughly and polishing with a cloth wheel or chamois.

You use liquid lacquer by cutting the end of the cable carefully and perpendicularly, dipping the end of the cable into the solution, and letting it dry for some 10 to 15 minutes. Edmund Scientific has the lacquer and sanding materials for their Crofon optical-fiber cables.

ADDING THE END FITTINGS

These are held in the cable by a close-pressure connection that may use one of the following techniques; heat shrinking plastic wrappers, metal pressure connectors, or gluing with a light transparent epoxy. You might want to add lenses to the ends of the cables. You can also add filters of various types to limit the type of light (frequency or wavelength) propagated through your fibers. Light sources that have been used with Crofon optical fibers range from Tensor lights to an ordinary light bulb, to lasers and LEDs. I caution again about the light: Be careful, especially with laser light. Don't take chances with your eyes. Use smoked goggles. Don't look directly into the light source, or allow anyone else to do so. Use any other precautions you can find out about to prevent damage to your eyes or those of your family, friends, and colleagues.

Finally, Fig. 8-15 shows the acceptance angle of light for a Crofon cable optical-fiber. This is interesting in that the acceptance angle is much larger than we have heretofore been led to believe it might be. Not only is the light-acceptance angle as shown, but the illuminating angle for light emerging from the fiber and the visual seeing angle are the same as that shown. It's a good wide view and makes light input relatively easy.

LIGHT DEVICES FOR COMPUTERS
ARE NOT LIMITED TO FIBEROPTICS

The search and research goes on looking for better and better mousetraps. A report from the Hughes Aircraft Corp. says their research department has developed a new all-optical logic device which is bipolar. It consists of four reflecting surfaces and a slab of nonlinear gallium-arsenide. It provides high speed switching in a unit that is impervious to any random or constant electrical interference. This device flip-flops reliability with a switching time as little as 3 nanoseconds and switch energies under 100 microjoules. They want this unit to come near the theoretical switching limit of 10 gigahertz/second so it can be used more effectively in computers, flight control systems, and signal processors. A research effort is being made to further reduce its size so it can be contained on an optical integrated circuit chip.

What could a bipolar switching device be used for? It operates on a light ray or rays. Imagine a window shade which can be opened and closed a billion times a second. Previous research has produced some devices (typically crystal) which can block the transmission of light rays when a given amount of electric potential is applied to them. When such a current is removed, this device permits passage of the light rays. For all practical purposes, this is a bipolar switch. This switch can be held in either state (on or off) for as long as the applied potential exists or does not exist. It is a controlled and reliable bipolar device. The Hughes device may not be just that, but it may act in the same manner. And the search goes on for better mousetraps.

USING OPTICAL FIBER CABLES IN AUTOMOBILES

Of course you are familiar with this concept. You simply terminate a cable near a taillight and run the cable (along with other wiring in a larger cable) to above the rear window. Then you put a cover on the tiny compartment with two holes in it so the ends of the optical cable can go into each hole. When viewed from the front seat (via the rear-view mirror) you can clearly see the little bright dots which show you that the taillights are on, or that the taillights are off. Activating your turn signal makes one little light blink so you know that the system is working.

You are probably familiar with the front fender light dots which show you when you have parking lights on, regular headlight beams on, and/or the distant illuminating beams on. This is a handy little extension above the front fenders over the wheels, to show you that your headlight beams are working properly. Fiberoptic cables connect these to the proper light element in the front of the car. You do not see any cables showing, nor do you have to worry about electrical problems or deterioration. The fiberoptic cables need no more energy than the light which goes into them, but they do produce results and are handy to have.

At least one company, Hitachi, has done some study on the effectiveness of replacing the electrical cabling system in a car with an optical fiber system. It would be capable of translating the current-voltage demands from one area to another. They suggested that laser light emitters might be used to transmit untold numbers of digital signals through optical fibers to suitable diode-type receivers. Their purpose would be to monitor functions and operations within the car itself and present this information for display or use for better automatic adjustment of the various devices that make the car run.

Their researches used some LEDs that produced light in the infrared region and sent it through very small diameter optical fibers (1 mm diameter) to suitable light receiving diodes. They are of the firm belief that the entire automotive electrical system might be replaced with such a system. The gains would be better efficiency, lighter weight, probably less cost in the long run, and less interference generation and susceptibility. The fact that computerization of automotive engine adjustments is increasing and being made more autonomous implies that reduction of electrical interference is a primary requirement for the complex computer devices to operate reliably and satisfactorily. In the future, such devices may permit automatic steering and control of your automobile, once it is on a properly instrumented highway. How nice! Just jump into your car, set the computerized control for office, and sleep, read the paper, or listen to the morning news. And finally, the car stops in its prescribed parking slot at the office.

MILITARY CONSIDERATIONS
OF OPTICAL-FIBER COMMUNICATIONS AND CONTROL

"Communications silence!" This expression indicates how important it is to be secure in the emission of electromagnetic radiations from ships, aircraft, tanks, vehicles, or other components of the military operational establishment. To transmit any information by normal radiation techniques could result in telling the enemy just where you are and provide the point-source information the enemy needs to accurately steer an electromagnetically-guided homing-missile to you. On shipboard, optical fibers can be a base communications system or connecting system between all the ship's elements for communications, control, computation, assessment and all other required functions. This connecting system, while having all the advantages of providing the necessary functions and information, does not radiate.

A LIGHT-OPERATED COMPUTER

At Ohio State University in 1981 an all-optical digital computer was demonstrated. This computer used a Hughes liquid-crystal light valve. This device accepts optical images (light rays) and replicates them on a completely separate light beam from an arc lamp or laser. The technology is similar to that used in the fabrication and operation of liquid-crystal watches, but the important thing is that optical equivalents of electronic logic gates and flip-flops have been constructed with these light valves. They perform in much the same manner as do the transistor logic and gate circuits of the electronic computers. In an optical computer, the computing operations are said to be much faster, the computer more immune to interference and, perhaps more reliable. This light computer uses photons (bursts of light energy) in the same manner that electronic computer uses electrons. Perhaps connections to the various sections of the computer is accomplished by optical-fiber connectors and cables. That also seems logical. Thus we may yet see a revolution in computer development in the years ahead. Optical systems have the main advantages of security (non radiation), and wide bandwidths, which could mean a much higher rate of data transference than would ever be possible with wires and electronics as we now know it.

MODULATING A LIGHT BEAM

At the Naval Post Graduate School of Monterey California, an optical communications system was constructed with a pulsed gallium-arsenide injection laser used for the trasmitter. It operated at a repetition frequency of 8 to 15 kHz using a pulse-position modulation concept to convey intelligence. A PIN diode was used in the receiver. The pulse position modulation was used to transfer information as well as to trigger the laser so the light beams produced had the same pulse pattern.

REMOTE CONTROL USING A LIGHT BEAM

Of course, someone would investigate the use of light beams as a means of conveying commands to a remote receiver for remote control of various devices and operations. The use of a laser as the transmitting source for this kind of experiment has been accomplished. The receiver (in order to be equivalent of a broadband conventional super-het) has to be one which looks everywhere for the incoming light flashes that represent commands. The receiver has a 360° field of vision. Such a remote control system, without isotropically radiating energy, is secure and in most directions silent. It can be used where no other type of control system can be employed. In the reports it was said that a continuous wave helium-neon laser served as a transmitter and that its output was suitably modulated to convey the command instructions.

In the field of information storage, it is amazing to learn that optical-storage systems offer an almost unbelievable capacity to retain digital data. A single optical disc is recorded with a special laser beam. It is the size of a long playing record and is capable of storing billions of digital data bits. Just how much data it stores depends upon the size and spacing of the laser produced holes in the disc, and in the type of coding used. From 500 million to 50 billion bit can be stored on a single side of such an optical-storage disc.

Some of the advantages of optical storage turn out to be the longer life of the disc as compared to the electromagnetic type of disc-storage. But there are also disadvantages. One of these is the inability to erase and change data on a given disc. It is said that the error rate should not be less than 1 bit in 100,000 — a pretty high factor. Those who use computers say that this rate is not good enough. They want only one error in 10 trillion bits!

Of course optically written data can be written on a credit card or other flat instrument used by the general public. The information stored is not only secure, but also may contain much more information than you might imagine can be contained in such a small area or space. The mechanics by which data is stored is in small pits or holes which are laser produced. This produces a different reflectivity to an impinging light ray than the surrounding material. Thus, if there is a good reflection, this might mean a one, while a pit, causing a much less bright reflection, might mean a zero. A pit or hole may be only one micrometer in diameter; that takes a small size laser beam with some good optics to get the scanning job done. One method supposes that the laser is modulated by a pulses input signal so it creates the pits and blank spaces between them in precisely the same relationship as the modulating signal information.

To read the data requires some doing. First there must be an appropriate light source, a laser with a pinpoint beam and fiberoptic cables to convey the output to a beam splitter, which will send a reference beam split to the disc to be reflected back. The second half of the beam is used to generate a reference signal so that the electronics knows whether the reflected beam is of full intensity (meaning no hole or pit) or is partially absorbed in the hole or pit so it has a lesser intensity. These two extremes of reflection can then be interpreted by the electronics detector and comparison circuits as being ones or zeros.

Of course, there must be a tracking system which causes the laser beam to follow the tracks of data on the recording. That is another complex electronics system, but one which has been developed and works satisfactorily. Also, you then can imagine the complexity of the system to go in to any portion of that disc to extract a particular piece of data. Drexler Technology makes a material for optical recordings called Drexon. RCA is also utilizing this field of optical storage, as are other respected companies such as 3M, Thompson-CSF in France, and IBM. This area of optical-storage development is growing and it will cause some changes in our concepts of computers and stereos and what they can do. Just imagine writing 450 million bits per second. One day maybe we'll find out that our own human brains are too limited and too slow in operation—what do you think?

USING OPTICAL FIBERS IN DIRECTIONAL REFERENCES

It was not too long ago that the old mechanical gyroscope was the standard reference system in guided missiles, aircraft, ships, and submarines. It is still used because it has proven to be reliable, rugged, and accurate. But new developments (better mousetraps) are always forthcoming! Modern technology now acclaims the laser gyro as a rotational rate sensor system which can be used to roll-stabilize missiles, aircraft, and (using computers with good memories) determine positions and directions.

The laser gyroscope operates on the principle that you can split a coherent light beam. If you send it along two different paths, a plane perpendicular to the direction of rotation of some body, then the light beam in the path moving in the direction of rotation will move faster than that which moves along a path opposing the direction of rotation. How do you know this? Measure the phase or frequency of the two beams after they have traversed the paths, and the phase or frequency is different. Of course, it is assumed that you start with a single beam which is in phase and split it to make up the two beams used (so that one acts as a reference to the other).

After much effort, researchers found that the device required the light to travel a long distance which translates into increased size, not exactly what they wanted. They looked to optical fibers, whose elements could be wound up on spools to make long paths. Since the fibers are small in diameter (and wind up into very small spools) they got the small sizes they needed. Again, the optical fibers assume a most important role in a very scientific device.

Everyone was pleased at the results demonstrated by Victor Valie and Dick W. Shorthill of the University of Utah who first demonstrated the practical construction and operation of such a device using optical fibers. These gyro units have a very slow drift rate (meaning the shifting of the reference) of about one degree per hour and this compares to some mechanical gyroscope's drift rate of up to 5° per hour in some cases. So, with the small size and accuracy available what is the trade-off? Some say it is the difficulty of getting the laser light beam into the optical fibers since the laser beam is a relatively large diameter and the fibers are relatively small in diameter. Also, the optical fibers still have some accuracy problems. The optical-fiber gyroscopes do not function as well as the ordinary, larger laser gyroscopes using the same principal of operation. Some say that the optical-fiber gyros are in the category of good mechanical gyros but improvements are on the way.

There is a need for much experimentation and development in optical-fiber connectors, light sources, transmitters and receivers, polarizers, and phase splitters. Much development is underway and will continue to produce optical fibers with little loss, wide windows and monomode operation to reduce dispersion losses. Just as we went from hard-wiring in electronics circuits to the printed circuits and then to the integrated circuits—so we now progress to optical-fiber connected elements and the use of light as the active element.

USING PREFOCUSED LEDS

Some firms are making prefocused LEDs for use with optical fibers. One such firm is the Honeywell Opto-electronics Division of Dallas, Texas. The lens becomes an integral part of the glass or plastic envelope of a LED as shown in Fig. 8-16. These plastic LEDs are low in cost, and can be used with optical fiber cables that might link together the home computers, TV sets and cable television distribution systems, and computer peripherals.

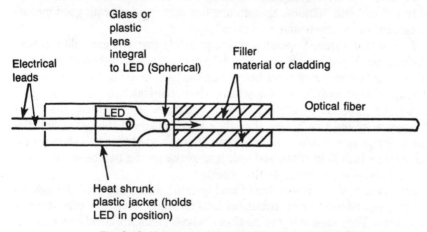

Fig. 8-16. Using a prefocused LED with an optical fiber.

In the manufacture of LEDs using a plastic covering, a lens is made right into the end of the unit and it may be spherical in shape. This is a somewhat different shape than you might have thought of as a lens. However, it is not unusual. As mentioned previously, light rays are an extension of the electromagnetic spectrum and the rays behave much like radio, radar, and other ultrahigh frequency radiations. The military has used focusing in their spherical Luneberg Lens antenna systems and there is no reason why the small glass bead of spherical shape would not focus light rays in the same manner in which a Luneberg lens will focus radar waves. (By the way, the concentration point of light in a LED unit, when focused, is called the sweet spot.)

MORE ON THE ROTARY MOTION SENSOR USING OPTICAL FIBERS

A rotary motion sensor which employs and endless closed-loop, multiple-turn optical fiber interferometer. Coherent laser radiations are coupled into the (optical) fiber interferometer in both a clockwise and a counterclockwise direction. The energy is permitted to traverse the entire light-waveguide many times to increase the accumulative phase shift which will occur due to the multiple transversals. A method to get the laser signals into and out of the light waveguide has been devised. Also a means has been devised to generate a pair of reference signals which are then hetrodyned with the clockwise and counterclockwise sensor signals to produce a pair of frequency signals at a manageable frequency. The relative phase shift between the acoustic output frequencies and the circulating signals is the same; hence, they provide an extemely precise measurement of the rate of rotation of the device. In a preferred configuration, the means of coupling and generating a reference signal comprise of an acoustical waveguide for generating a required opto-acoustic grating across the paths of the coherent optical signal and the optical-fiber interferometer. The detection of phase differences between reference signals and sensor signals seem to require some kind of diffraction process.

The sensitivity of such an optical-fiber system to inertial rotations is adequate when the loop consists of 600 turns of single-mode fibers. In the systems tested, the name Sagnac was used as identification. Also much work was done on light amplifiers which could compensate for the losses in the optical fiber path. When such a system can accurately detect the very slow rotations of objects with respect to inertial space (which is assumed to be fixed in rigid coordinates) then the uses of such a rotational sensor will be increased many fold.

WHAT ABOUT AN OPTICAL DELAY LINE?

The delay line is often used in Early Warning Radar systems to preserve the frequency and phase relationships of rf signals of the microwave frequency range. Such delay lines have to meet very rigid requirements such as low insertion losses, wide range and bandwidths, and produce a nearly linear response. If we convert the microwave energy to light energy and then use an optical-fiber delay line we find that a spool of the optical-fiber materials is practical. It takes time for the energy to traverse such a spool and thus delays are obtained. What is needed is a single-mode type of fiber and a high speed photodetector and rf amplifier. Studies have been made of such delay lines operating in the 4.0 to 6.5 gigahertz range. Direction modulation of the light signal is accomplished with the use of an injection laser.

OPTICAL-FIBER COUPLING DEVICES

A patent has been granted which discloses a portable unit that can couple optical fibers together so energy will transfer among them. This unit consists of an insulative block to which the optical fibers are attached. A piezoelectric block is attached to the fibers also and connected to a source of energy. When electricity is applied to the piezoelectric element, an arc is developed across the electrodes in such a way that the optical fibers are heated and caused to flow together, joining them. The coupling is solid and tight and as good a couple as can possibly be made. It is better than pressure-joining the polished flat ends of pairs of optical fibers.

AN OPTICAL-FIBER MECHANICAL SENSOR

The U.S. Navy Department has invented a mechanical sensor that uses an optical-fiber in a sensory role. This concept uses an optical fiber to modulate the intensity of light passing through an optical fiber to modulate the intensity of light passing through an optical fiber in response to mechanical motion in the form of linear position, rotational rate, or strain on the fiber. The light from a suitable source is send down an optical-fiber path through a birefringent modulatable element which is positioned between cross-polarizers. Then the light goes on to a suitable diode detector. The birefringence of the modulating element is susceptible to the influence of an external source of pressure such as a mechanical stress movement or strain. It is also susceptible to change due to an applied magnetic field that can be generated by the device being monitored or sensed. This type of external force causes a rotation of the light in the modulating element. Thus it encodes the information when there is a mechanical change, producing a digital signal that then can be detected and decoded by suitable electronics and computers.

When light passes through an optical fiber, it can be weakened or dispersed by small anomalies in the fiber such as nonsmooth surface effects of a winding spool. Subjecting the optical fiber to any physical stress or strain causes some change in its light-conductivity to take place. You can detect this change in light transmission and relate it by some mathematical algorithm to the physical changes in the sensed element. This yields a good sensor.

Earlier, his text described birefringence. You may recall what it is: If you take a crystal, such as calcite $CaCO_3$ and lay it over some printed words on a page, the image you see appears double. This is due to the formation of two beams of reflected light with different polarizations with respect to one another. One ray of light is refracted in the normal manner and obey Snell's law of refraction. The other beam of reflected light does not obey this law, but passes through the crystal as though it were not there. Thus, two distinct beams are formed by this phenomena. By the same token, when light is passed into such a crystal, two beams emerge: one is a direct ray and one is a refracted ray. They occupy two distinct positions in space. The fact that two images or beams are produced is called the phenomena or birefringence.

Let us continue with the idea of an optical fiber used as a sensor. In connection with previous developments, the Navy Department has produced a mechanical shutter effect for use with light rays and optical fibers. A mechanically actuated piezoelectrically operated, fiberoptic connected light valve (shutter) can be used as a mechanical motion (pressure) sensor or displacement sensor. The piezoelectric element responds to pressure by generating an electrical signal that is applied to a Kerr cell containing nitrobenzene. This electrical signal then makes the nitrobenzene transparent (it was not transparent to light previously). Hence the light rays can easily pass through this Kerr-cell and through the conducting optical fibers to a light detection receiver where the light intensity is then converted into a proportional electrical signal of the digital form, or a coded signal of the digital form.

Of course, optical fibers are used in hydrophones as sensors. An acoustic wave detector formed from an optical-fiber coil is placed in such a device. Since acoustic waves are pressure waves, then changes in pressure on the coil can be detected and so a hydrophone detector unit is formed.

DEVELOPMENTS NECESSARY FOR OPTICAL FIBER VIABILITY

Perhaps one of the most important areas of development to increase the use of the optical fibers is that of developing the light source in small integrated packages and developing light detection-conversion receivers in very small packages. Now they can be integrated into the ends of optical fibers. Already, light sources are being constructed using small lenses, lasers and LEDs. They are being complimented by receiver units which can produce linear outputs over wide bandwidths. Of course it is desirable to have optical fibers operate single mode, if possible, to reduce dispersion losses. Integrated optoelectronic devices, using silicon substrates, and single and multi-mode optical-fiber coupling systems are needed. New injection lasers, studies to provide more understanding of infrared physics, photon excitation, optical-fiber linkages, and computer-controlled laser systems as sources are all under investigation at present.

OPTICAL FIBER CONNECTORS

As you have probably already considered, a problem could develop using optical fibers if they are rigidly connected to terminal devices. This presents a dilemna. If you don't have the optical fiber tightly held in a terminal connector, it might slip or work loose and reduce the transmission of light rays. If it is very tight and secure, then any twisting or bending of the fiber might cause a fracture and render the light-guide inoperative.

A government invention offers one solution. This invention relates to an apparatus for attaching an optical fiber cable to a terminal using a snap-ring to achieve a rotatable connection wherein one jaw of the support device supports the terminal and the second jaw supports the ferrule. An intermediate third jaw supports and compresses the snap-ring. With the snap-ring initially mounted on the terminal, closing the jaws together compresses the snap-ring so that the terminal can be fully inserted into the ferrule until the snap-ring is captured within the ferrule thus completing the connection.

OPTICAL FIBER MEASUREMENT DEVICES

As with radar microwave guides, a system of measurments is necessary to determine the efficiency of the light guide when using optical fibers. An invention which describes how a light guide can be measured using a combination of optical properties has been developed by the Department of Energy in Washington D.C. A polarized light pulse is injected into one end of the optical fiber. Then the reflections from discontinuities, which are not polarized are adjusted to impinge upon a light detector. The polarized incident light pulses cannot reach such a detector. Using the intensity of the detected reflections, the transmissibility of the light fiber can be measured.

USE OF OPTICAL FIBERS WITH PUNCHED CARDS

Of course, using optical fibers with punched cards is now almost old hat. Groups of optical fibers are arranged so that they cover the total area over which a card might have holes punched in its surface. The card is presented to the fibers with a backlight source. When there is a hole, light passes to the optical fiber. Where there is no hole, light is blocked. The resulting light-dark combinations can then be converted into an alphanumeric display, print out, or whatever. The interesting concept here is that a very complex and close-reading grid can be established using optical fiber ends. The data (light pulses) can be converted into electrical binary pulses some distance away in some electronic device and used for whatever purpose is desired (even a check of hole-punching accuracy).

CONSIDERATION OF THE PHOTODIODE RECEIVING ELEMENTS

In GaAsP APDs have been fabricated with uniform high-speed gains, higher than 42, and with quantum efficiencies of over 70 percent at 1.28 micrometers. At high gains there are increased leakage currents which can produce excessive shot-noise and gain-saturation. Much effort has been spent to try to reduce this bad effect. One method is to develop guard-ring APDs but the results have not been conclusive. Much more effort is underway to improve the light detector capabilities, speed of response, bandwidths, and linearity. They are essential to the further application of light rays in computing systems. Noise generation as indicated here is a real problem, with high gains.

SOME FANCIFUL USES OF OPTICAL FIBERS

Since optical fibers can be used as sensors and can transmit light which falls on their ends, it seems that it should be possible to devise a checker set which could use optical fibers going to the various squares. This would convert the light-dark sensed changes to electrical signals that could be used by a small computing element to determine chess movements. Imagine a checkerboard that has optical fibers terminating under each square. When that square is covered by a checker, the light from the room will be blocked off so no light passes. When the square is uncovered, by movement of the checker, then the light falls on that end of the optical fiber and so passes through to its detector and so on.

There is no doubt that other games might be devised to use ambient light, or light in beams or bulbs, to cause activation of various functions. One such device is the light gun and target game. Optical fibers can connect the various target elements to detectors so that as pulses of focused light fall on them, a panel could light up showing the score.

Sometimes it is convenient to have the light-detector right at the point of light-entrance to the device, and sometimes it is not. When it is useful to have the computer element, coding element, or decoding element some distance away, optical fibers are ideal to convey the light from the entrance point to the device.

USING OPTICAL FIBERS IN MEASUREMENT OF ROTARY SPEED AND ANGULAR POSITION

NASA devised two optical sensors, one a 360° rotary encoder and the second a tachometer to operate with a light source and a remotely located detector. The light source and the detectors were coupled to passive scanning heads through 3.65 meter fiberoptic cables. The devices were used on certain types of engines that NASA had interest in. This is another example of where the optical fibers might be fruitfully used where the environment precludes the possibility of having the sensors directly coupled to the detectors and they must be located some distance away. Not only was the reflected light obtained through optical fibers, but the source light was also sent to the sensor through these optical cables. This type of two-way channeling of light rays presents unlimited possibilities.

CONSIDERATION OF OPTICAL FIBERS UNDER DYNAMIC EFFECTS

The attenuation of both low and high numerical aperture single-mode fibers has been found insensitive to either low frequency dynamic tension or dynamic twisting in a study by ITT Electro-Optical Processes Division. The evaluation of the fibers with respect to dynamic bending showed that low NA single mode fibers were attenuated during the bending phase while the high NA fibers were unaffected. Coating materials having refractive indices either higher or lower than the fused quartz were applied to the single-mode fibers and were found to have no effect on the dynamic attenuation characteristics. No effects resulting from dynamic interactions were found; all attenuation effects observed during the dynamic testing were explainable in terms of twists, tension, or bending.

USING OPTICAL FIBERS IN DANGEROUS SITUATIONS

There are some situations in which the use of electronic or electrical sensors could be very dangerous. One such situation is that of monitoring high-explosives or highly inflammable materials. Optical fibers with a light source presented to the sensor and the reflected light returned to a remote detector by means of optical fibers solve this kind of problem. The reliability of this type of sensing and the light transmission systems used is very high.

USE OF OPTICAL FIBERS IN COMPUTING ROLES

Sperry Research produced a report which described the development of a fail-safe optical data bus system that uses several terminals. The data bus is in a serially-operated type of system. The system consists of a master terminal containing a CPU (Central Processing Unit), LED sources, a photodiode detector, and several remote send/receive types of terminals. The transmission system is composed of multi-mode monofibers and data can be impressed on or tapped off at each terminal of the system. The purpose of the investigation was to develop a concept for the system so that it would continue to operate even if there was a power failure at one or more of the remote terminals.

After considering many types of terminals, an angle collimated terminal and a mirror terminal were constructed. Both were compatible with Corning low-loss multimode optical fibers with NA (numerical apertures) of about .15. Both devices used about 85-micrometer thick crystals $LiTaO_3$ to control the light flow in the optical fibers using the electro-optic effect of these crystals. The mirror-terminal was selected for implementation to meet the research goals. It was proved that the system consisting of three mirrors on terminals met the objectives and proved to be of a fail-safe nature.

AIRBORNE USES OF OPTICAL FIBERS

There is no doubt that a means to prevent complex airborne electronics systems from becoming susceptible to electromagnetic interference (EMI) or electro magnetic pulses (EMP) purposefully generated is important. With the amount of computer activity used on board most modern-day airplanes, the miles of wires connecting the elements furnish ideal pick-up points for stray and undesired radiation. Optical fibers have no such susceptibility and their use makes the systems just that much more reliable and effective.

Airborne avionics systems are moving toward the uses of party-line multiplex data buses for the increasing number of digital signals used in modern aircraft. The opto-electronics developments now provide a data bus interface system. Much work is still needed to provide suitable couplers for optical fiber cables. It has been necessary to provide two-way transmission of the light-data into and out of the main data-buses. Each station on the bus must be able to communicate with every other station on the bus. In some cases a nine-port radial coupler has been used and pluggable optical interfaces have been employed with cable lengths of up to 30 feet. A maximum bus length in earlier trials was up to 100 feet. When the system used unipolar Manchester coding on the optical bus with a data rate of 10 megabits/second, a worst-case error rate of 1 part in one million bits was experienced.

But the critical area has been found to be in the couplings which permit two-way data transmission. Pulse compression techniques have been used to reduce the losses at such terminations and other techniques are being constantly devised. Some types of couplers under investigation are the so-called star planar type couplers. Both theoretical and experimental characterization of 8 by 8 transmission couplers have been made with waveguide structures fabricated on a Nd-rich glass substrate using an ion-exchange process. The structure patterns were formed on the substrate by means of a photolithography process.

Interfacing using switching techniques may employ an electro-optical switch which can couple any one of four waveguide channels. This may be selectively induced by the application of an electric field between intersecting strips of electrodes placed adjacent to the top surface of the slab and adjacent to the bottom surface. These electrode strips comprise primary electrodes to which constant voltages can be applied to induce optical paths partially through the slab. Also, switching electrodes are used which are vertically placed and electrically isolated from the primary electrodes. To these an alternating voltage is applied that causes a proper and desired optical path through the slab to be accomplished.

USING THE FAST FOURIER TRANSFORM FOR PULSE ANALYSIS

The Fast Fourier Transform, adaptable to modern computers, can be used in analysis of the responses of optical fibers and their detectors. The National Bureau of Standards has an automatic pulse measurement system to measure the pulse responses of optical communications components and to compute their impulse and their frequency responses. For example, the measurement of the properties of two glass fibers (fiberoptics) and an avalance photodiode using both a pulse (GaAs) laser wavelength of 0.9 micrometer, and a mode-locked NdYag laser wavelength of 1.06 micrometers has been accomplished. The measurements were performed in the time domain. In order to get the frequency domain data, the Fast Fourier Transforms were used with suitable computers. The impulse-type responses were obtained by deconvolution. This reminds me of the early days when I tried to evaluate the responses of some hi-fi amplifiers I had developed. I used square wave pulse inputs of various frequencies to test their responses.

STUDYING THE LOSSES IN OPTICAL-FIBER CONNECTIONS

McDonnell Douglas research department has been studying the optical interference losses between the transmitter-to-receiver interface systems. The pulse dispersion effects, rise times of light transmitters and receivers, types of optical fibers available, and detector sensitivity were all considered. Also, many bussing systems such as the tee, star, and hybrid systems were carefully analyzed. The matter of using a single fiber versus multi-fiber cables also was evaluated. They were interested in the use of optical fibers on board a space shuttle. They considered a data bus on a space shuttle and then made an optical analog, using optical fibers, which might replace the hard-wire system. System tests using a 9-port star data bus system for an optical-fiber bundle considered losses, connector difficulties and efficiencies, data rate, and also the susceptibility of such a system to EMI (electromagnetic interference) in the range of 200 GHz to 10 GHz. Also, they tested such a system as to its vulnerability or nonvulnerability to lightning. It turned out that optical fibers were ideal in every instance and could be used on all future space shuttles. Thus, if that type of engineering is your ultimate goal, remember to seriously study the use of optical-fibers and all the elements associated with them!

THE JOSEPHSON JUNCTION

A Josephson chip can multiply an 8 bit by 12 bit array in a few nanoseconds 10^{-9}, which is much faster than current conventional transistor-semiconductor circuits. Since speed and interference immunity are primary goals in future computer developments, this device becomes of primary importance. It is a switch arrangement some 10 times or more faster than the fastest type switching circuits currently in use in integrated circuitry. The switch depends on a super conducting concept for its basic operation.

One concept of the switch, the Bell Labs Josephson-Atto-Weber Switch, called JAWS, is shown in Fig. 8-17. In the off condition, a current circulates in the triangular loop to the off line and ground. In the on state, the loop junction resistance becomes infinite and diverts the input current to the output line. The switch essentially acts like a short circuit element across a line. When the junction elements are at a low conducting state the current ignores them, and goes on the output. When the junctions are in a highly conducting state, a

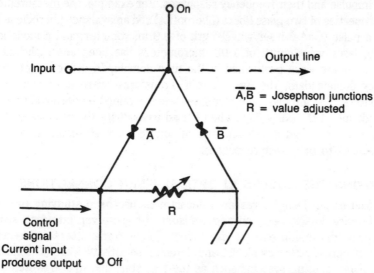

Fig. 8-17. The Josephson switch (courtesy of Bell Labs).

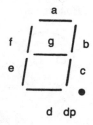

Fig. 8-18. A seven-section LED.   Digit display

control current must be applied as shown. This development is in its most elementary state at present but it is so important that you can expect to hear much about it in the future. It takes its place along with fiberoptics as a means to increase the speed of operation of various devices such as computers and it will help reduce their size.

One further idea needs to be expressed with respect to this switch. It needs to be very, very cold to operate correctly. In its tests, the Bell Labs put it into a Deward flask of liquid helium which keeps the super-conducting junctions at a proper temperature. A circuit will be able to perform many operations and present them in parallel in its output when many of these switches are formed into an array.

SOME RANDOM OPTICAL CIRCUITS

After reading about or studying new concepts such as optical fibers and the circuitry associated with them, you may want to get some hands-on experience. To this end, I conclude this chapter with some random circuits that use LEDs and/or other optics in their operation. One of the most common uses for the light-emitting diode is that of displaying numbers. One such unit for doing this is the Radio Shack 276-060 LED 9-digit display block. It uses GaAsP LEDs for display and it can easily be interfaced with TTL, DTL, or MOS units. It is illustrated in Fig. 8-18. Of course these displays can be obtained in a single unit or a multi-unit.

Let's examine the connections for some LEDs. The first in Fig. 8-19 is a subminiature red LED lamp-type with a diffused lens. You apply a small voltage (1.6 volts) and it glows. Figure 8-20 shows a single LED of the jumbo size. It uses 1.75 volts for its operation. There are many other types of LEDs that might be just what you are looking for. Ask about other types at your radio parts store.

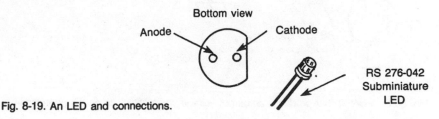

Fig. 8-19. An LED and connections.

Bottom view

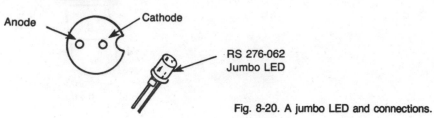

Anode   Cathode

RS 276-062
Jumbo LED

Fig. 8-20. A jumbo LED and connections.

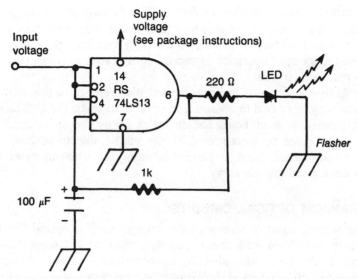

Supply
voltage
(see package instructions)

Input
voltage

1
14
2
RS
4
74LS13
7
6

220 Ω

LED

Flasher

1k

+
100 μF
−

Fig. 8-21. A simple LED flasher circuit.

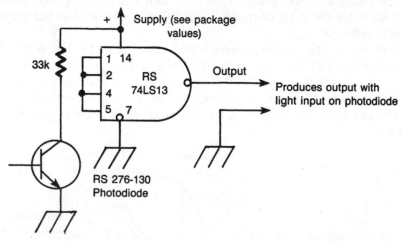

+
Supply (see package
values)

33k

1 14
2
RS
4
74LS13
5 7

Output

Produces output with
light input on photodiode

RS 276-130
Photodiode

Fig. 8-22. A circuit that produces an output voltage when light falls on the photodiode.

214

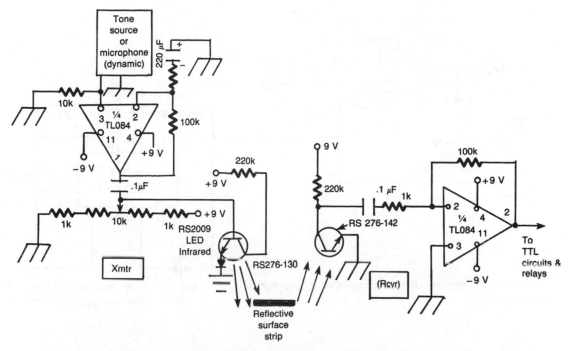

Fig. 8-23. One type of circuit for robotic path control, an alarm system, or a communications system using infrared light rays.

You might want to make a flasher unit with a reasonably high rate of flashing. One such circuit is shown in Fig. 8-21. Figure 8-22 illustrates a receiving circuit which can produce an output when a light falls on the input diode. Of course all these circuits may have to be experimented with and component values adjusted to get the kind of operation you want.

Here are some circuits that are fascinating both for their possibilities and use in some exotic applications, and for their illustrative indications of how opto-electronic circuits are put together. Figure 8-23 shows a simple infrared transmitter and receiver which might be used in a path-finding application of a robot. You might have to do some experimenting with the resistor values. When you purchase the parts they will no doubt have some circuit diagrams as part of the package. These might be used effectively.

One LED transmitter or flasher can be built around a 555 integrated circuit as shown in Fig. 8-24. Also, shown is a LM 3909 unit in a very simple connection to produce some light flashes. A pulsing infrared light might be used to give a robot some idea of direction—that is, it might be made to home on the flashes. Or it might be used to warn the machine of possible confinement areas or obstacles or trouble spots which it should avoid. Control of the light flashes can be used to send various control codes by infrared light. I leave that idea for your development.

The opto-coupler is an ideal way to isolate one sensitive circuit from

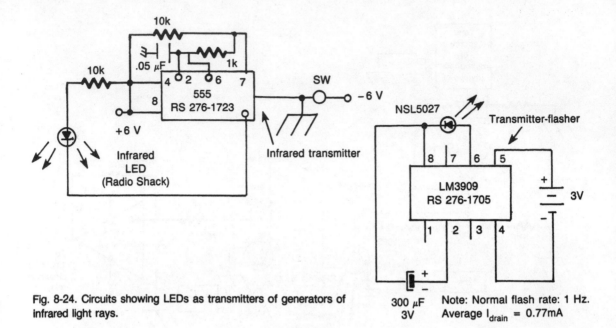

Fig. 8-24. Circuits showing LEDs as transmitters of generators of infrared light rays.

Note: Normal flash rate: 1 Hz.
Average I_{drain} = 0.77mA

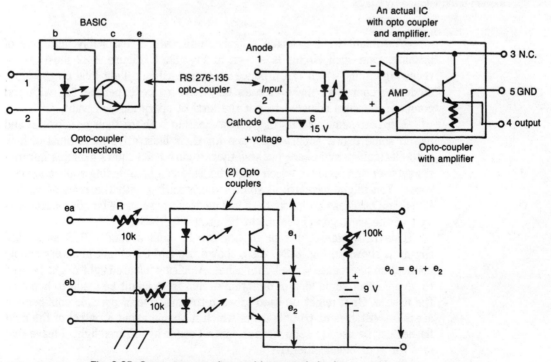

Fig. 8-25. Some opto-couplers and integrated circuits.

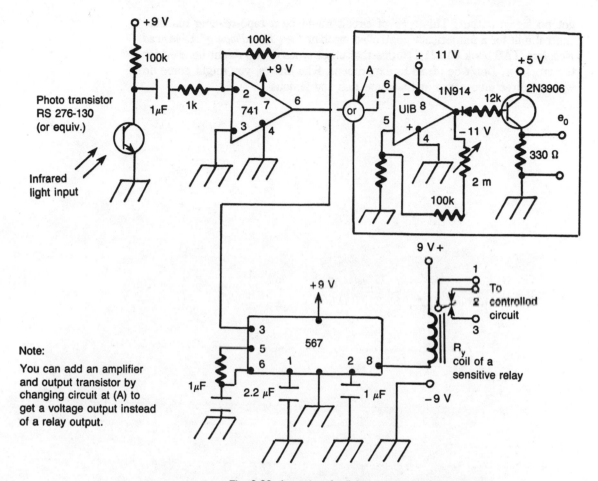

Photo transistor
RS 276-130
(or equiv.)

Infrared
light input

Note:

You can add an amplifier
and output transistor by
changing circuit at (A) to
get a voltage output instead
of a relay output.

Fig. 8-26. A receiver for light-control of another circuit.

another (for example, a power control circuit from an input circuit). Figure 8-25 shows the basics of such circuitry and some integrated circuit schematics which may be useful to you. There are many types of opto-couplers. Some use triacs for control. If you want to make your own coupler, there is no reason why you cannot use an LED transmitter coupled into an optical fiber, and connect a receiver photodiode to the other end to convert the light to current-voltage. Two opto-couplers can be used to sum some input voltages as shown (called mixing). It can have application when you need to have two sources (sensors) provide some inputs and then not have a reaction until both sensors say it is time to do so, or until one sensor has such a large output that you need to have some action taken by the (robotic) machine.

A receiver circuit for light rays or light pulses is shown in Fig. 8-26. In this circuit a sensitive relay is caused to close when light falls on a phototransistor receiver to produce a signal which can be amplified and presented at the output of the LM 239 amplifier. If you block off the light path, you might get no signal output. This type of circuit might be a tape-reading machine control unit for a numerically controlled machine (see *Handbook of Advanced Robotics,* TAB book #1421). Notice that other types of ICs might be used for the amplifier. Don't be afraid to experiment. Who knows, you might come up with a better mousetrap and become rich and famous!

Chapter 9

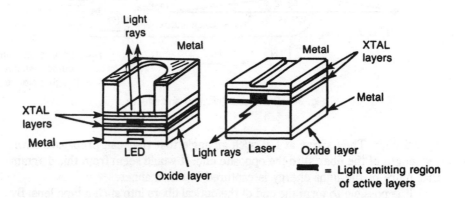

Lenses and Fiberoptics

W E HAVE DISCUSSED REFRACTION, REFLECTION, AND THE BASIC COMPOSITION
of light. When such rays pass through mediums with various indices of
refraction they can be bent or refracted. A light wave that front hits upon a
power type of surface (a mirror comes to mind) has light rays reflected from
that surface at specific angles. The angle of reflection equals the angles of
incidence. Now let's examine the focusing of light rays through various types of
lenses. Lenses are used with light sources which, in turn, pass the light rays
into optical fibers.

FORMING LENSES IN ENDS OF OPTICAL FIBERS

When you consider that the focusing effect of a lens is due to different
indices of refraction between air and the material of the lens, then you may
wonder if it would be possible to form a lens out of the optical-fiber material
itself. It would actually make no difference whether the end of the fiber were
the end of a bundle of fibers or a single fiber. If there are many fibers in a
bundle you might assume that various light rays would go through one fiber and
other rays would go through other fibers.

Figure 9-1 shows a source of light energy at 0 and two rays. One goes
directly into the glass through the radius of curvature of the end r terminating
as shown at I meeting with the refracted ray which hits the curved surface.
Then it is refracted to also meet at I. The fact that the two meet at I in a lens
effect is important in lens considerations. The two rays are now inside the

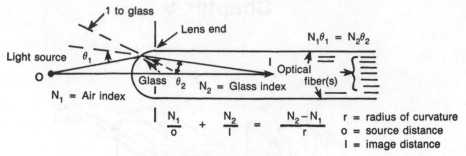

Fig. 9-1. The optical-fiber end-lens effect.

optical fiber. They arrive at I at such an angle that they can propagate down the length of the fiber(s) to the opposite end. It would seem from this diagram that more of the light energy is captured in this manner.

It is possible to form the end of the optical fibers into such a type lens. By heating to the proper temperature, the glass becomes plastic and can form or be formed into such a lens. Some skill is required and very careful temperature control is needed so the glass won't run or ball, but it can and has been done. This could be important in coupling applications and capture applications.

The reverse kind of situation might be possible by making a lens out of the terminal end of the fibers. Physics shows us that with proper refractive qualities of the fibers and medium beyond the end of the fiber(s) that a refraction opposite of the input effect shown in Fig. 9-1 can take place. The source becomes the focal point and the I point of that figure must be a cross-over of the rays inside the glass material so that this becomes an inside source origin of the rays.

In a manner of thinking, the thin-lens effect occurs at the end of the optical fiber. One curved side of the thin lens is not really formed but is present as illustrated in Fig. 9-2.

There are other types of lenses and effects that can be obtained using materials with different indices of refraction. What is important is that focusing and small source rays are requirements in optical fiber systems, even when the fibers are in bundles. This means that you cannot handle optical fibers or optical fiber cables like you would handle hard wires. The end connections, joining elements, coupling elements, and feeding and receiving elements have to be carefully designed. Lens effects must always be considered. You just don't clamp the ends of two fibers together as you might join a pair of wires and expect the transmission of the energy to be equivalent. Much more care and caution is necessary in the handling of the optical fibers.

HOW MANY CHANNELS CAN THERE BE IN A SEGMENT OF THE LIGHT SPECTRUM?

One of the most important concepts related to the use of light as a transmission medium and the use of optical fibers to guide these light rays is that of gaining more channels. You can mix data, TV, radio and telephonic communication on wires of the two pair variety and the coaxial variety, but there is not enough room or channels to carry all the information required. More channels are needed and continue to be needed as our need for data transmission and information increases. How to get the extra channels is the question.

Let us consider a normal coaxial cable transmission system that can transmit frequencies up to several gigahertz. The bandwidth or channel-width for good telephonic communications is about 3,000 Hz/second (3×10^3). Radio broadcast channels require about 10 kHz for their channel spacing. In the microwave region, channels are some 250 to 300 kHz in width. To find the number of channels available in a given span of frequencies, simply take the width of the channel to be used, plus its separation channel, and divide this into the width of the space available. The answer is the number of channels that can fit into that spacing. An example: If a two wire system can transmit frequencies from dc (direct current) up to 1 MHz (one million hertz/second) and good telephonic communication requires 3,000 hertz channel width, then

$$\text{Channels} = \frac{1 \times 10^6}{3 \times 10^3} = 333.333333 \ldots$$

A little over 333 channels per hard wire pair! Obviously these aren't enough channels to be practical, so consider a coaxial cable which can handle up to 1,000 megahertz/second. Now our figures become:

$$\text{Channels} = \frac{1000 \times 10^6}{3 \times 10^3} = 333.33333 \ldots \times 10^3$$

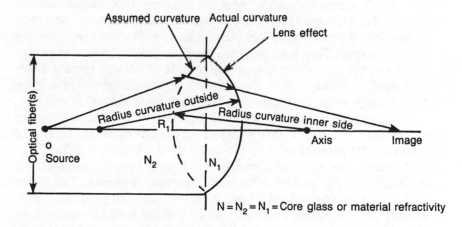

$$\frac{1}{o} + \frac{1}{i} = (N-1)\left(\frac{1}{R_2} - \frac{1}{R_1}\right)$$

Fig. 9-2. The thin-lens concept at the end of an optical fiber.

221

or a thousand times more than the wire pair. The line is also probably smaller even though terminal receiving and sending equipment and couplings may be more complex.

These computations assume a simultaneous transmission of all channels. Actually, time-division multiplexing or frequency-division multiplexing increases the traffic handling capability, so we consider optical fibers with their wide band capability. Let us assume for this imaginary example that we can transmit visible light in the region of 450 to 650 Å over such a cable. An optical fiber might be used singly. Consider a visible light band of 200 angstroms, which represent a space available channel width of:

$$\text{Hertz} = \frac{2.997924 \times 10^8}{200 \times 10^{-10}} = 14,989.62 \times 10^{18}$$

For the communications problem requiring a 3×10^3 bandwidth, you find:

$$\text{Channels} = \frac{14989.62 \times 10^{18}}{3 \times 10^3} = 4996.54 \times 10^{15}$$

If you write 15 zeros after that number you'll see why fiberoptics is important, at least in the telephonic sense!

SOME PROBLEMS WITH CHANNELING USING OPTICAL FIBERS

When discussing the number of channels that might be obtained over optical fibers using light rays, you need to at least be aware of the other effects which tend to limit these channels. In the first place, light is not exactly pure. That is, it isn't exactly on a single line of the spectrum. Lasers come closer to having this single line spectrum than any other light source, but you can expect some unexpected effects which might tend to broaden the light spectrum even from a laser source. Then it encompasses more frequencies (instead of a single one) and we have a narrower channel spectrum to fill because the non-usable channel itself is wider. Fewer channels can be accommodated in a given frequency range if the nonusable portion of the source channel is wider.

Then there are other effects, which include intermodulation distortion (the by-product frequencies of input frequencies or pulse rates) and intermodal dispersions of frequencies called by the nonpurity of the index of refraction of the optical-fiber line or cable throughout its length. I have mentioned that pulses might be distorted if a line has much intermodal dispersion. This causes the pulses to be so rounded and delayed in transmission that problems exist in recovering the intelligence from them, so the channel spectrum required for good intelligent transmission may not be calculated as simply as indicated in the elementary example.

Think of channeling in another way. You have assumed an analog type of frequency, a continuous wave tone, voice, or data pattern. Now consider the effect of pulses and pulse rates. To transmit a pulse requires a wide bandwidth:

$$bandwidth = \frac{1}{pulse\ duration}$$

to use an old radar formula. This is the minimum bandwidth required to recover a reasonably shaped pulse. Then add to this the pulse rate itself. Assume that the pulses are transmitted at half a million hertz. That represents another frequency component into the line and its harmonics. Assuming a varying pulse rate (depending upon the type of intelligence represented), there are a multiple of equivalent frequencies present. You can think of a system in which time division multiplexing restricts the pulse rates to only one during an interval of time being present in the line so the multiple by-products are not problems. That, of course, is a possibility.

SOME PATENTS CONCERNING OPTICAL FIBERS

It is always interesting to examine patent information about optical fibers. As time goes on there will be more and more of them as the use of these fascinating elements increases. One patent is directed to a security device that detects tampering with a secured inclosure. A fiberoptic bundle is looped through a closure and secured at opposite ends of the bundle to a snap-together connector. An intermediate length of the fiberoptic bundle surrounds the snap-together connector preventing access to its locking mechanism unless the fibers are cut. After installation, light is passed through the fiberoptics and a particular pattern is generated at the viewing end of the connector. Tampering with the inclosure causes the individual fiberoptics to be disturbed so that subsequent viewing of the particular pattern indicates something different than that which would be seen if no tampering were done, so you knows that the lock has been tampered with.

This brings to mind a possibility which might come about one of these days. A window glass might be made so that an optical fiber strand curls back and forth throughout its area. This would not be visible ordinarily, nor would it impede the normal transmission of light through the glass when viewed from the front or rear. But an invisible light pattern could be sent through this channel. If the glass were broken in any way, the light-to-electronics receiver would then sound an alarm, much like we now get with hard-wire type conductors, or the vibration-sensing bugs around and on the windows.

A three dimensional memory has been invented having a vastly increased storage capacity, using optical fibers. The memory block has a matrix of cylindrical cavities. Each has a fiberoptic light-guide embedded inside. Each light-guide is composed of a cylindrical core having a first index of refraction and a cladding surrounding the core that has a second index of refraction. This second index of refraction is smaller than the first. There are a number of spaced deformations formed at the cladding-core interface that permit the light to leak out laterally. This light contains the memorized elements.

Another interesting area is the use of optical fibers in a hydraulic diagonstic monitoring system. This system warns of failures in the hydraulic system components by using on-board sensors that continuously monitor various failure indicating parameters. The monitoring system consists of three basic types of sensors: analog, discrete, and fiberoptics. The sensors feed information to a self-contained, centrally located display panel through the various interface circuits that are easily replaced and readily accessible to maintenance personnel. Also the display panel indicators that might fail are monitored, indicating a malfunction when there actually is no malfunction.

There is an invention that is a remotely controlled solid-state relay with a fiber-optic input. The relay is powered by the circuit being controlled and thus uses no power of its own. A trigger circuit powered from the ac line is actuated by back-biased PIN diodes. The trigger circuit controls a switch that is a pair of silicon-controlled rectifiers in series with the ac source and the load. The light input through the optical fiber triggers the circuit, which in turn, permits the rectifiers to function and thus supply power to the controlled device or circuit.

DEVELOPMENT OF FIBER-OPTIC DEVICES

One report concerned with coherent fiber-optic testing techniques fits well into our discussion. Any development has to be proved, so it is worth considering the following from the symposium of Photographic Instrumentation Engineers.

"We have found that reliance on theoretical models or incomplete manufacturing data is not adequate for predicting the results of combining optical and electro-optical components into systems. The complex imaging characteristics of each component must be accurately known if reliable predictions are to be expected. In-house testing is desirable when possible. Some components have well established evaluation procedures and others such as coherent fiberoptics (as of 1979) require test techniques, some of which are well established and others which must be devised, to find out if suitability is accomplished and if complete characterization (or definition of operational characteristics is to be found). This is especially true where changes in magnification are a feature of the bundle (of fiberoptics). The technique used for testing can include image rotation, magnification, image size variation, distortion, shear, transmission efficiency, transmission variations, transmission defects, flatness and acceptance angle."

When you consider the types of tests as specified here, you begin to realize what it means to ensure that a given batch of optical fibers will be usable in some particular application. Continuous testing of this order of magnitude may then be a mandatory requirement in a manufacturing plant. Also, the use of new and improved testing devices and methods are of constant interest in this fast growing field of science and engineering.

Some definitions of devices used with optical fibers are of value here. An optical access coupler and a duplex coupler have some special meanings. Access couplers are for use with multi-terminal communications systems. The duplex coupler is a device which allows a single fiber to be used for data transmissions in two directions. Such devices may consist of special grooves of different depths and widths that have been etched into silicon along natural planes. This way you can position fibers of different diameters properly and securely in the grooves, and provide couplers, reflective silvered surfaces for proper reflections. Optical communications signals pass through smaller diameter transmission lines to a larger line through this kind of device. The light going from the smaller line to the larger line may be reflected by these reflective surfaces, and varying them permits adding the intelligence desired.

In some coupler applications it is necessary to couple energy that comes from a light waveguide into a more ordinary rf waveguide. This can be accomplished with a single fiber held in a capillary tube and positioned by a micro-positioner so that the greatest light output is obtained. The fiber is then secured to the waveguide by means of epoxy at one end of the waveguide channel. The two elements are then positioned in the capillary tube in such a way that they cannot be rotated.

Reflective surfaces can be used to channel and guide the light energy from one end of one optical fiber cable into another cable. The positioning of these reflective surfaces must be exact, and they must be small. They must be rigidly positioned along the fiber ends so that no movement or distortion of position is possible. In considering such an arrangement, you can visualize how a multiple of combinations of light energy might be summed together in such an arrangement. Also you can imagine how an input line might convey the light energy to a given reflective position and that energy then subdivided through the reflective position so that many individual strands can be energized with the incoming rays.

While considering the use of optical fibers in transmission systems, the mechanical strength of the fibers becomes an issue. The cable must be handled in the field and so it must have some strength. In one effort to increase the strength of these optical fibers a ceramic coating is applied over the optical fiber core. It is an inorganic material impervious to moisture and chemically corrosive materials. It is not the cladding material, but is applied over the cladding in the drawing process.

SHOCK WAVE EFFECTS IN OPTICAL FIBERS

Although the optical fiber is impervious to electromagnetic effects and can be used in many applications where large magnetic fields exist, it has been found in some applications that shock waves in solid materials can turn the fiber opaque and prevent its transmission of light rays. Thus, if the fiber may be subjected to these shock waves, you have to take precautions to protect them. There are some beneficial effects from this phenomena as pointed out in some studies. Due to the fact that the shock wave does make the fiber opaque, it can be used as a sensing and measuring device to generate a signal when the shock wave is present and absent. In some research applications this timing capability can be important. Of course other timing applications can generate from this light-no-light transmission concept. This can be even to the point of severance of the fiber due to some mechanical or chemical or other action that interrupts the light transmission at a specific time.

When the fiber is accurately positioned with respect to a receiver so that the light rays emanating create a signal in the receiver then a disturbance may misalign the fiber and receiver, and the signal will vary or cease at that instant. Thus a marker of the time of the event is generated.

A SHORT REFRESHER COURSE IN OPTICS AND LENSES

What is light? Light is a radiant energy which affects the eye and enables us to see. Light can be visible to us and can be invisible to us. Light moves at a specific and unvarying speed. Light rays are parallel, but light is a component of the electromagnetic spectrum, and James Clerk Maxwell proved this to be the case. This is important because you can then consider that light is simply the far end of the radio-radar-microwave spectrum. A chart such as shown in Fig. 9-3 is helpful to visualize this concept.

The spectrum is logarithmic, of course. And the visible portion of the spectrum is from about 450×10^{-9} to about 650×10^{-9} meters. The word millimicron is often used in optics with the abbreviation ($m\mu$). It is interesting that the upper frequency of human vision is the violet-blue region and the lower frequency is in the red region. The eye is most sensitive in the green-yellow region. Now you know why lots of signs are colored yellow with black, or dark lettering.

Of course, all people do not see the same colors with the same vividness, but we won't worry about that here. Our optics systems, being electronic and electromagnetic can detect whatever range of frequencies (colors) we design them to detect. Notice that optics systems may be somewhat color blind if they aren't designed properly. This means they may not transmit the light rays of some colors and frequencies. The limits of human sight are from about 4300 Å to about 7000 Å.

Computers and other electronic devices can be used to enhance the things people look at (pictures or TV screens etc.) so that we can see colors which we might not ordinarily see. Note this on the next weather-news broadcast you see on TV. Most such programs use color enhancement to show wet and dry areas. Fiberoptics requires the generation of certain colors and frequencies of

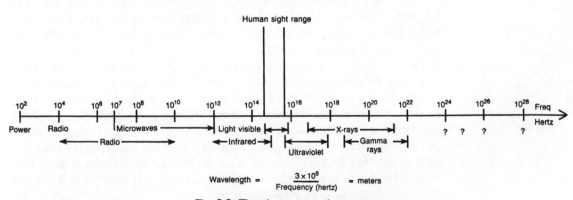

Fig. 9-3. The electromagnetic spectrum.

light for proper transmission through fiberoptic channels. You can imagine some of the problems if that light color is not constant, or at least relatively so. Light sources are very important. Some LEDs do not generate constant frequency outputs. Some lasers require feedback and automatic adjustment of electrical energy to maintain a relatively constant power and frequency output. Since the frequency is on a logarithmic scale, a very small change results in a considerable effect. Much effort is needed to maintain the constant frequency of the light source.

WHITE LIGHT

What you perceive as white light is not a single wavelength of light at all. Sunlight is considered to be white light and it is composed of all wavelengths of light energy. You can artificially mix various wavelengths of light energy to form what the eye sees as white light. In a way, this is like white noise in electronics systems. The noise is composed of all frequencies of noise mixed together.

SOME CONSIDERATIONS OF LIGHT ENERGY

If you consider light rays to be simply an extension of the electromagnetic spectrum, you are in agreement with most modern physics texts. Under this consideration, the energy in the light rays is described by the familiar poynting vector of a radiated wave:

$$\hat{S} = \frac{1}{u_0}\hat{E} \times \hat{B}$$

This is illustrated in Fig. 9-4. The resultant $\hat{S}$ vector is the cross-product of the

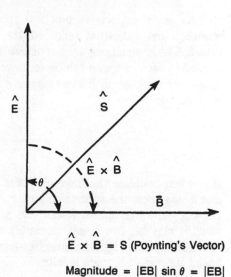

$$\hat{E} \times \hat{B} = S \text{ (Poynting's Vector)}$$

$$\text{Magnitude} = |EB| \sin \theta = |EB|$$

Fig. 9-4. The vector cross-product gives Poynting's vector.

227

electric vector $\hat{E}$ and the magnetic vector $\hat{B}$. This shows the direction of wave motion, and the vector cross-point mathematically gives the magnitude of this wave.

What has worried some physicists is that the light rays also exhibit a momentum (pressure) effect. We know this because we can measure the pressure on some types of surfaces—recall the small four-leafed axle in a vacuum bottle with its silvered leaves? When placed in the sunlight so the leaves reflect the sun's rays the axle spins in a direction to cause the leaves to revolve away from the sun's rays. The speed of rotation is dependent upon the brightness of the sun's rays. In the glass-vacuum container, there is no air friction—or at least it is minimized—which otherwise would slow down the rotation. In modern space platforms this pressure effect of sunlight may have considerable importance. Some studies have actually considered a sun-sail which might be extended from the body to gain pressure to cause motion in space. Fascinating idea, isn't it?

It has been determined by experiment (after Clerk Maxwell said the effect would be evident) that the momentum given to a body could be calculated by some equations that are very simple in expression. (I am always amazed at how most of the world's important equations are simple three letter combinations such as $E = Mc^2$, $E = IR$, $F = Ma$, etc.) Here again is the same type of expression:

Momentum when ray is totally absorbed by body: $E/c = m = p$

E is light energy, c is speed of light, and m is the momentum (also written P).

Momentum when energy is totally reflected: $m = \dfrac{2E}{c} = P$

As you'd expect, in practical applications where energy is somewhat absorbed and somewhat reflected, the value lies somewhere between these values. These equations are important in calculating, for example, the size sail needed to cause a space vehicle to move at some value of acceleration through space, assuming no air friction magnetic friction or viscosity. Since acceleration is given by:

$$a = \frac{F}{m} = \frac{\text{Force}}{\text{mass}}$$

If you can calculate the force produced by the sun's rays on a given size sail, and if you know the space ship's mass, you can determine the acceleration of the body through space. Recall that a constant acceleration, no matter how small it may be, can result ultimately in almost the speed of light! It cannot reach the speed of light because Einstein showed that when a body reaches this speed its mass becomes infinite.

THE QUANTUM THEORY AND THE PHOTOELECTRIC CONCEPT

It used to be that because some substances emitted electrons (called photo-electrons or photons) the electromagnetic theory wasn't considered usable to explain this effect. The quantum theory could be used for such explanations. There were other effects about this photon emission which furrowed brows (the velocity of emission is not influenced by the intensity of the light rays). Emissions increase with the frequency of the light radiations. The amount of emission does very with intensity of radiation impinging on the target.

So, how was all this explained? It was explained by assuming that the light energy consisted of small energy packets called quanta, or photons, and these had different energy values, depending on the light frequency or wavelength. Again, that simple three-letter type of equation described the energy content:

$$E = hf$$

E = energy, h Plank's constant, and f = light frequency in Hertz. The energy related to the electrons in order to satisfy Einstein's equation became:

$$\frac{1}{2} mv^2 = hf - W = hf - hf_0$$

Where m = mass, v = velocity, W = work function or energy required to remove one electron, hf = lowest value of hertz at which emission takes place, and hf_0 = frequency of light applied.

THE PHOTOELECTRIC EFFECT

Of course, the photoelectric effect is what interests us mostly here. By impinging a certain type of light on a certain kind of substance under certain conditions, the substance gives off electrons of a type measureable as electric current. Isn't that what you want in optical fiber transmitter-receiver systems? Millikan won a Nobel prize for his work in connection with this effect in 1923. It was found that this photoelectric effect was a kind of surface phenomena and that if the surface was contaminated with grease, dirt, oxide films, or other such inhibitors, then the photoelectric effect was vastly reduced or nonexistent. This is very important in the manufacture of photoelectric devices used as receivers and converters of light rays to electric currents.

There are some contradictions when considering the wave theory of Maxwell to explain light wave radiations. The quantum theory seems to fill the gaps and permits a more unified and acceptable explanations of these contradictions. Some of these are: Wave theory says kinetic energy of photoelectrons should increase with an increase in light intensity. Experiment shows this not to be so. The kinetic energy remains independent of the light intensity. Wave theory also says there should be a photoeffect no matter what the frequency of the light rays may be. This is not found to be true. There is a cut-off frequency which is the lower limit of the effect. There should be some delay in the emission of photoelectrons when the light rays may be. This is not found to be true. There is a cut-off frequency which is the lower limit of the effect. There should be some delay in the emission of photoelectrons when the light rays are weak. This is not so. There is no delay no matter how weak the rays may be.

Max Planck believed that light traveled through space as an electromagnetic wave, but Einstein showed that light had to travel as small particles. Einstein's expression (proved in experiments by Millikan) seems to be most valid when considering the photoelectric effect.

This analysis of the use of optical fibers and light shows that the wave theory is used to analyze the propagation of the light rays through the fiber. However, you would then consider the quantum theory when analyzing the devices which convert those light rays back into electrical currents for use by computers, and other such types of equipment. You might want to investigate these theories in much more detail than have presented here. Just remember that, currently, scientists consider light to have a dual nature (behaving like a wave in some circumstances and like a particle in other circumstances). The particles are called photons.

ON THE DUALISM OF LIGHT RAYS

Louis deBroglie got involved with matter waves. He reasoned that since light had a dual nature (wave and particle) that perhaps matter should have some wave properties also. He stated that the wavelength of matter could be found by:

$$\lambda = h/p$$

where λ = wavelength of light, p = momentum, and h = Planck's constant. He said that for matter, p would be the momentum of a particle of matter.

Huygens (1680) had a theory about light waves but it was not accepted by the scientific community because he couldn't figure out what the wavelength of light had to be. In 1800 Thomas Young was able to perform this calculation and made the concept and theory of Huygens much more acceptable. With much work many scientists then got into the act bringing forth wave mechanics of Erwin Schrödinger, the probabilities of particle-position existence of Max Born, and Heisenberg's uncertainty principle (1927). It was Niels Bohr who discussed the complementary nature of waves and particles, and he said they didn't contradict in theory.

THE CONCEPT OF LIGHT SOURCES AND ILLUMINATION

If you have a light source at some point in the room that the light rays go out in all directions (assuming no reflectors or focusing) so that each point of a spherical surface in space surrounding the light source gets energy from the source at the rate of:

230

$$E = \frac{I}{y} \pi r^2$$

$\pi = 3.1416$, I = intensity of illumination in ergs, and p = distance to sphere surface.

There is the equation of a sphere divided into the available energy in ergs, so the energy is distributed all around the sphere interior as you should expect it to be distributed. It is very interesting to note that this same kind of approach is taken when calculating the energy from a radio or radar antenna that is nondirectional. The equations are the same as for light.

Some texts say that if you have two sources and two spheres you can adjust the size of the spheres or the distance of the source from the sphere surface until your eyes tell you that the energy falling on that surface is the same for each of the two sources. Actually, this may not be true as your eyes may not tell you the truth, but visually the illumination is the same. At that time there may be a difference in the size of the spheres of the radius of the spheres and so on. But, when you get this condition, you tend to say that the energy from the source is the same for both of these resulting spheres. They each have the same illumination as the other. The energy in the illumination can be expressed as a simple fraction:

$$\text{Energy} = \frac{I}{r^2}$$

where r = sphere distance, and I = intensity.

This is the inverse square law of radio-radar and microwaves with which we are all familiar. The energy falls off proportional to the square of the distance (r) over which it travels.

Since your eyes may lie to you about the intensity of illumination, and this fact was well known by earlier scientists, they decided to do some experiments to remove this difficulty. They devised some experiments (involving the Lummer-Brodhum photometer) that still required visual comparisons but were more accurate. Today photoelectronic devices (light meters) are used which easily and accurately measure the intensity of light from any type of source, even reflected light rays.

Why all the worry about equal intensities of light? Well, if you are going to create a light source for an optical fiber system, and you have some idea (by other calculations) of losses for that frequency of light, then you might work backwards and calculate the intensity of the light source in order to get the required energy through the fiberoptic system, over a given distance, with a given type of optical fiber. It is nice to be able to use calculators and computers to work out all the values you need to have to make such a system work correctly. That way you don't have to do endless experiments, which are costly and time-consuming, to get the answer. You work out the mathematics of the situation and components and problems and then you conduct some experiments which are valid and require the minimum of time, cost, and energy, to prove the validity of your conclusions.

Sometimes you need light of a given color—which means a given wavelength—to match an optical-fiber system. Although light is composed of hundreds of shades of color (maybe thousands) there are fundamentally just six basic colors used that we can observe easily and effectively. These six colors are red, orange, yellow, green, blue, and violet.

White light is multi-colored light, so is it possible to separate some colors from the white light? The answer of course is yes, and you can use a prism to do so. White light into one side of a prism emerges as several beams due to the frequency diffraction deviation, or color separation, and wavelength refraction due to the different index of refraction for the different frequencies in the material that makes up the glass prism. Some light frequencies in the material that makes up the glass prism. Some light frequencies (colors) are refracted more than others, so the basic color spectrum appears. The basic color spectrum is the six colors mentioned, and Fig. 9-5 illustrates a normal type dispersion which is not to any scale but is just illustrative. You do not find the colors so markedly separated in an actual experiment; they tend to blend together. But the sketch does show how the different frequencies will be refracted one more than the next one and that is what we are interested in at the moment. It now seems that if you do not have a source of monochromatic (one frequency) light (such as obtained from a laser) then you might use a prism

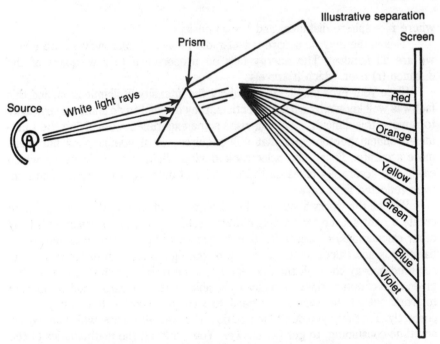

Fig. 9-5. A prism separates the colors of the spectrum.

and a white light source to get a given frequency (color) for optical fiber experiments. It would take some doing but it might be done. The word dispersion can mean spreading out the colors of light. So saying there is dispersion in an optical-fiber system and that this causes losses means that the various frequencies are separated in the fiber and some are much delayed or bounced around more than others. There are differences in prism materials. Some may absorb some colors and emit others and so on. Infrared is the invisible part of the spectrum beyond red, and ultraviolet is the invisible part of the spectrum beyond the violet. The wavelengths range from 800 Å to some 4500 Å, where 8000 is in the infrared region. The angstrom unit is 10^{-9} meter. You might study prisms and the light effects through them and be astonished at the results.

IN CONCLUSION

Light has a wave-particle duality and the electromagnetic wave nature of light is the factor which allows it to propogate through a wave guide of fiberoptic material. Lenses and filters are the controllers of light in an optical system. Understanding the fundamentals of optics, basic concepts of physics, and quantum mechanics lets you grasp the full potential of lasers and fiber optic cables. We can only dream of the next generation of communications available through the use of these powerful tools.

As science continues in its quest of knowledge, the challenge will be for technology to apply this newly-found information for the good of mankind. The weekend researcher and at-home science buff have key roles in this progression. Your imagination today may be the major breakthrough of tomorrow.

Index

shock wave effects in, 225
sources for, 197-198
splicing of, 127, 128, 161, 162
strength of, 96, 160, 161, 176
testing of, 178
transmission capability testing for, 142, 143, 172
transmission windows for, 145
unguided invisible light in, 148
using cylindral lens to couple lasers to, 174
waveguides for, 175
optical path length, 59
optical radar systems, 75, 76
optical system data transmissions, 67
time delays in, 74
optical window, 158
opto-electronic circuitry, 195
optocoupler, 131-133, 215, 216
optoisolator, 146-148
ordinary (O) ray, 39, 41

P

parabolic mirror, 125
patents, 223
Pauli Exclusion principle, 21
peak transmission, 45
peak wavelength, 45
penta-prisms, 43
phase change, 117
phi prime, 27
photocells, 195
photodiodes, 135, 209
photoelectric effect, 5, 229
photographic applications, 46, 134, 153
photometers, 231
photons, 5, 7, 9, 17, 86
photophone, 139
photoresistor element, 195
physics, 1-23
pi meson, 17
piezoelectric crystals, 65
PIN diodes, 93
laser communications systems using, 79
quantum-efficiency curves for, 94
Planck's constant, 5, 14, 229, 230
Planck, Max, 230
plane mirrors, 32
planetary-satellite concepts, 6
plano lens, 29
Pockels effect, 118
polarization, 35, 36, 40, 41, 59, 70, 112, 115, 143
circular, 22

crystals and prisms for, 40
devices and methods for, 38
dichroism, 39
phenomena of, 144
proper plane of, 55
rotation of light during, 118
shifting in, 118
symbols used in, 42
potential energy, 6, 7
Poynting vector, 37, 38, 115, 116, 227
pressure effect, 228
principal quantum number, 15
prisms, 110
color spectrum separation through, 232
laser beam use of, 43
polarization phenomena and, 40, 41, 144
reflection and refraction in, 89
protons, 3
pulse analysis, fast Fourier transform for, 211
pulse coding, 140, 141
optical delay lines for, 175
pulse quaternary modulation (PQM), 72
pulse-position modulation (PPM), 80
pulse-switching system, 50
pulses, 140, 142
pumping circuit, 48
punched cards, use of fiberoptics and, 208
pyro detectors, 56

Q

quanta, 2, 5
quantum chromodynamics, 22
quantum electrodynamic theory, 13, 19, 22, 229
quantum-efficiency curves, 94
quarks, 17-20

R

radar, 1, 85
optical, 75, 76
wavelengths of, 86
radiation hardened optical-fiber system, 181
radiation, wavelengths and photon energy of, 17
Raman effect, 168
ranging, 60, 61
rare gas halide (excimer) laser, 56
rate detector, 82
Rayleigh scattering, 168
receivers, 114, 127, 129

recovery time, 94
reflection, 32, 35, 88, 99
reflection mirrors, 24
focusing with, 31
refraction, 27, 35, 88, 99
double, 112, 144
index of, 27
varying indices of, 100
remote-control systems, 78, 202
retardation, 41, 42, 60
robotics, fiber optics in, 104, 136, 152-153, 184-193
rotary speed sensor, 209
ruby lasers, 2, 56, 64, 96
Rydberg atoms, 13, 14

S

safety, 96, 174
sampling, 130, 141
sapphire crystals, 2
satellites, 6
Schlieren-quality glass, 43
security systems, 120, 223
semiconductor diode, 173
sensors, 114, 115, 134-136, 205-209, 223
serial-access couplers, 181
shells (see atomic structure), 4
shifting (see polarization), 118
shock wave effects, 225
Shorthill, Dick W., 204
signal transference, 140
silicon marker laser, 56
silicon mirror, 53
single-mode fibers, 101, 106, 117, 142, 182
Smekal-Raman effect, 168
Snell's law, 27, 99
Snell, Willebrord, 99
solid-state lasers, 9, 10, 46, 48
sonar, 145
space experiments, 74
spectra, 15
spectroscopy, 17
spectrum, optical, 110, 226, 232
spheric lens, 31
spherical aberration, 29
splicing optical fibers, 127, 128, 161, 162
Stark effect, 66
step index, 117
Stokes lines, 168
strange-flavored quark, 18
strip diode lasers, 11
Sweig, George, 18
switches, 65
switching system, 49

T

tapping, waveguide, 176
tau, 22
TEA lasers, 49, 53, 56
telephone systems, 92, 93, 97, 98, 139
fiber optics and, 173
laser use in, 93
thin film polarizer (TFP) beam splitting, 60
time delays in optical transmissions, 74
Townes, Charles H., 1
tracking systems, 72, 120. 148
transmission, single mode of, 117
transmitters, 114, 129
temperature control laser light stabilization in, 130
transverse waves, 115
tuning, 70
piezoelectric crystals and, 65

U

up-flavored quark, 18
uranium (U-235 and U-238), 12

V

Valie, Victor, 204
vectors, 37, 38, 115, 227
velocity, 6, 7, 86
video-laser recorder, 81

W

Wallaston prism, 115
wave front propagation, 115
wave theory, 229
waveguides, 175, 176
wavelengths, 102, 230
waveplates, 58-60
white light, composition of, 227
wide-band distribution systems (see cable television), 169
window
frequencies of, 10, 44
optical, 145, 158

X

x-rays, Bragg's law and, 66, 67

Y

YAG lasers, 83
Young's experiment, 68, 110
Yang, C.N., 22
Young, Thomas, 116, 230

7388